Heat

Author
Alan J. Hirsch

Program Consultant
Marietta (Mars) Bloch

Australia • Canada • Denmark • Japan • Mexico • New Zealand • Philippines
Puerto Rico • Singapore • South Africa • Spain • United Kingdom • United States

Contents

Important safety information

Record observations or data

① Refer to numbered section in *Nelson Science & Technology 7/8 Skills Handbook*

Unit 2 Overview

You use and control heat every time you cook food, or change the temperature of the room you are in. Your body works hard to prevent or encourage heat transfer so your internal temperature is constant. The weather outside depends on heat transfer from the Sun to the atmosphere and water. The steel in the spoon you used for your last meal was made using huge quantities of heat. The electricity you use to dry your hair or drive your games may have been generated using heat from fossil fuels. In this unit, you will learn about the many ways we depend on and control heat.

Temperature and Heat

Temperature and heat are not exactly the same.

You will be able to:

- explain the difference between heat and temperature
- estimate and compare the temperatures of different objects
- investigate examples in nature of sensitivity to temperature
- record the temperature of a substance as it changes state and analyze the results
- identify factors that can affect the rate of temperature change

Heat Transfer

Heat transfers between objects of different temperatures.

You will be able to:

- observe and describe what happens to solids, liquids, and gases when they are heated
- describe how convection currents in Earth's atmosphere and oceans can affect weather everywhere
- choose appropriate materials when designing a device that conducts or insulates
- experiment with absorption of radiant energy
- compare the heat capacities of various substances

Producing and Using Heat

The sources and effects of heat are important when designing and building.

You will be able to:

- explain and demonstrate how insulating materials control heat transfer
- classify different sources of heat as renewable or nonrenewable
- describe how solar heating can be applied to reduce our need for other sources of heat
- recognize the causes and effects of heat pollution and describe ways to control it
- identify and explain how feedback devices are used to control heating systems

Design Challenge

You will be able to ...

demonstrate your learning by completing a Design Challenge.

Devices That Control or Use Heat

Through science and technology, people have learned to build various structures and devices to control or transfer heat to their advantage.

In this unit, you will be able to design and build

1 A Device That Delays Heat Transfer

Design a container that will keep a cold pop can as cool as possible from breakfast until lunch time.

2 A Swimming Pool Heated by the Sun

Design a device that uses only energy from the Sun to heat the water for a model swimming pool to 25°C.

3 A Greenhouse to Protect Plants

Design a greenhouse that will allow plants to grow even when outdoor temperatures are low.

To start your Design Challenge, see page 54.

Record your thoughts and design ideas for the Challenge when you see

Design Challenge

Getting Started

Thinking About Heat and Temperature

1. You know from experience that these onion rings are hot. After all, they were just deep-fried in very hot oil. But how will you know when they are cool enough to eat? Can you tell by touching them with your finger? Just how reliable is your sense of touch?

2. Hang gliders and many birds rely on a form of heat transfer. Rising warm air keeps them aloft. But why does the air rise? What is the source of the heat?

3 Electricity is produced at this generating station by burning natural gas. The heat released by the burning fuel is used to make steam, and the steam drives a turbine that creates electricity. The electricity is then transmitted through wires to our homes, where it may become a source of heat in appliances or even a furnace. At every step, heat escapes. Is this the best way to get energy from one place to another? What are the advantages and disadvantages of burning fuel? What other ways can you think of to heat your house that would be more sustainable?

Think about questions ❶, ❷, ❸. What other questions do you have about heat? As you progress through this unit, reflect on your answers and revise them based on what you have learned.

Try This Judging Temperatures (6A)

It is not always easy to judge temperatures by looking at a photograph. Is it any easier if you can feel a substance? In this activity you'll find out how well your hands can judge temperature.

Work with a partner to record your (9D) reactions. Then switch roles and let your partner try this.

- Fill three plastic bowls and place them in a line: one with hot water (but not too hot!), one with lukewarm water, and one with cold water.
- Place your left hand in the bowl of hot water and your right hand in the bowl of cold water for at least one minute.

1. Record what each hand senses.

- Place both hands in the bowl of lukewarm water.

2. Record what each hand senses.

- When your hands have returned to their normal temperature, touch several objects in your classroom. Do they feel warm or cold?

3. Record whether each object feels warm or cold.

4. What makes objects feel hot or cold?

SKILLS HANDBOOK: (6A) Obtaining Qualitative Data (9D) Working Together

SKILLS MENU
○ Questioning ○ Hypothesizing ○ Planning ● Conducting ● Recording ● Analyzing ● Communicating

Identifying Temperatures

As you have seen, it's difficult to judge exactly how hot or cold something is just by touching it. A better way is to find out the temperature of the object. For now, you can think of temperature as a measurement of how hot or how cold an object is. We usually measure temperature in degrees Celsius (°C). For example, the temperature of the palm of your hand is probably about 35°C.

In this investigation you will use your experience and cues from photographs to estimate temperatures in degrees Celsius. After you have arranged the temperatures in order from lowest to highest, you can evaluate your own arrangement.

Materials

- 28 index cards
- pencil

1. The oven when a pizza is cooking

Question

Can we estimate temperatures?

Hypothesis

From our own experience, we can create a scale of temperatures, using familiar objects and events as markers.

Experimental Design

Using your experience and deduction, you will assign a temperature to 14 objects or events.

2. Hottest day ever recorded on Earth's surface

3. Comfortable room temperature

Procedure

1 In a group, record items 1 to 14 on separate index cards.

(9D) **2** After discussion in your group, sort the cards into three piles: "low temperatures," "everyday temperatures," and "high temperatures."

3 Discuss which of the items in each pile of cards is the coldest and which is the hottest.
- Arrange the cards in each pile in order from coldest to hottest.

4 Make a set of temperature cards, one card for each of the temperatures (a to n) listed in **Table 1**.
- Match the temperature cards with item cards 1 to 14, making your best estimate of the temperature in each situation.

(a) Record the temperature that your group decided for each of items 1 to 14, listing them from the lowest to the highest.

Analysis

5 Your teacher will supply the actual temperatures for items 1 to 14. Record these temperatures on your item cards. Compare your group's estimate with the actual temperature for each item.

(a) Which temperatures were the easiest to predict? Why?

(b) Which temperatures were the hardest to predict? Why?

(c) Describe ways you could improve your skill at estimating temperatures.

SKILLS HANDBOOK: (9D) Working Together (4A) Research Skills

4. Boiling water

Table 1

a)	−273°C	**h)**	15 000 000°C
b)	−89°C	**i)**	37°C
c)	80°C	**j)**	160°C
d)	−10°C	**k)**	0°C
e)	100°C	**l)**	40°C
f)	20°C	**m)**	6000°C
g)	58°C	**n)**	7°C

Making Connections

1. How does knowing the predicted outdoor temperature help you plan an outdoor activity? Give examples for both summer and winter.

Exploring

2. (4A) Using a variety of electronic and other sources, research the Fahrenheit (°F) and Kelvin (K) scales. How are they the same as the Celsius temperature scale? How are they different? Where is each used? Summarize your information in a chart.

5. Lowest temperature possible

6. Air in a refrigerator

10. Hot tea

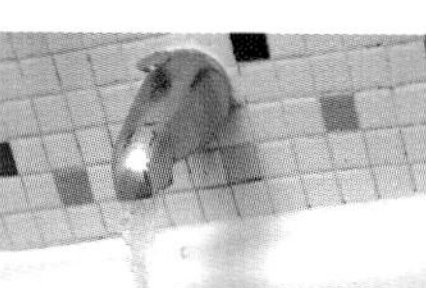

13. Comfortable bath water

14. Freezing water

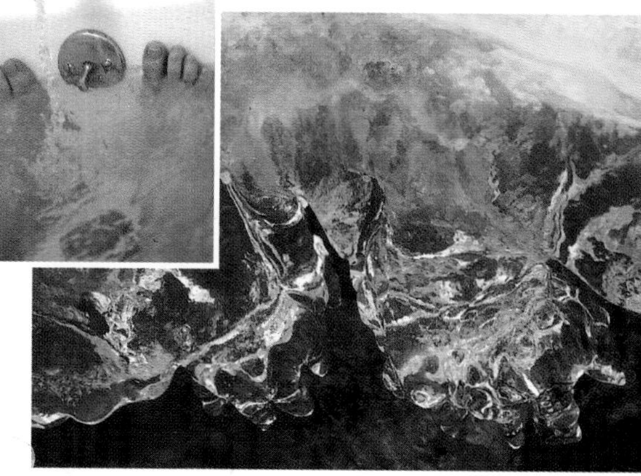

11. Healthy human being

12. Coldest weather ever recorded on Earth's surface

7. Ice cream

8. Interior of the Sun

9. Surface of the Sun

Design Challenge

Temperature is important in each of the Challenges.

(a) Estimate the temperature of cold pop from the refrigerator. Explain your answer.

(b) In the swimming pool challenge, you are asked to heat water to 25°C. Based on what you have learned, is this a good temperature for water in a swimming pool? Explain.

(c) People have a standard body temperature, but what about plants? Do pine trees or roses have a "body temperature"? Explain.

2.2 Inquiry Investigation

SKILLS MENU
- ○ Questioning
- ● Hypothesizing
- ○ Planning
- ● Conducting
- ● Recording
- ● Analyzing
- ● Communicating

Heating and Cooling

It is not wise to put a cold glass plate onto a hot stove. The plate could crack and break easily. All forms of matter change when they are heated or cooled. What happens to a plate as it is heated also happens to other materials, such as water and air. Learning about the effects of heating and cooling will help you understand how many things work, such as hot-air balloons and thermometers.

Materials

- apron
- safety goggles
- water
- food colouring
- Pyrex flask
- rubber stopper with inserted glass tubing
- hot plate
- retort stand
- clamp
- glass-marking pen
- ball and ring apparatus
- pulse glass

Question

What happens to a liquid, a solid, and a gas when each is heated or cooled?

Hypothesis

(2C) **1** Make a prediction for what you would observe for each type of substance, and write a hypothesis for your prediction.

Experimental Design

Water at room temperature, a ball and ring apparatus, and a pulse glass will be used to observe what happens to each substance when it is heated or cooled.

Do not attempt to remove glass tubing from the stopper. The glass may break.

Once you have turned on the hot plate, do not touch the surface of the plate.

Procedure Part 1: A Liquid

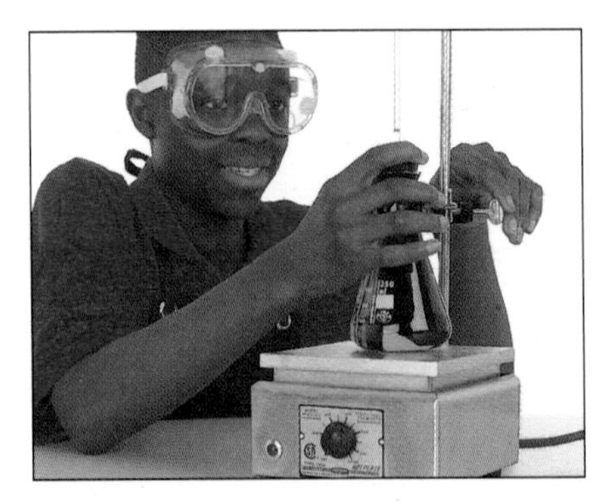

2 Fill the flask completely with room temperature water and add a few drops of food colouring.
- Insert the stopper with the glass tubing into the flask so that there is no air in the flask.
- Mark the water level in the glass tubing with a glass-marking pen.

(a) Why do you think you were asked to put food colouring in the water?

3 Put the flask on the hot plate and clamp it.
- (5C) Turn on the hot plate and slowly heat the water. Do not allow it to boil.
- Mark the water level in the glass tubing again when you have finished heating the water.

(a) What happened to the water level after the water was heated?

4 Turn off the hot plate. Allow the flask to cool and observe any change in the water level.

(a) Does the water level rise or fall during cooling?

SKILLS HANDBOOK: (2C) Predicting and Hypothesizing (5C) Using Other Scientific Equipment (9D) Working Together

Design Challenge

How might knowing that materials change size when they are heated or cooled affect your choice of materials for your Challenge design? How could you test materials to discover if they are appropriate?

Exploring

1. Do other liquids expand by the same amount when heated? Repeat the investigation using vegetable oil. Compare the results.

   Do not heat vegetable oil directly on the hot plate. When you are heating the liquid, place the flask in a hot-water bath.

2. How could you prepare your flask thermometer so it gives readings in degrees Celsius? (This process is called calibration.) For example, many thermometers use the temperature of ice water to calibrate 0°C and boiling water to calibrate 100°C. Explain how your calibrated thermometer could be used to measure the temperature of other liquids.

Part 2: A Solid

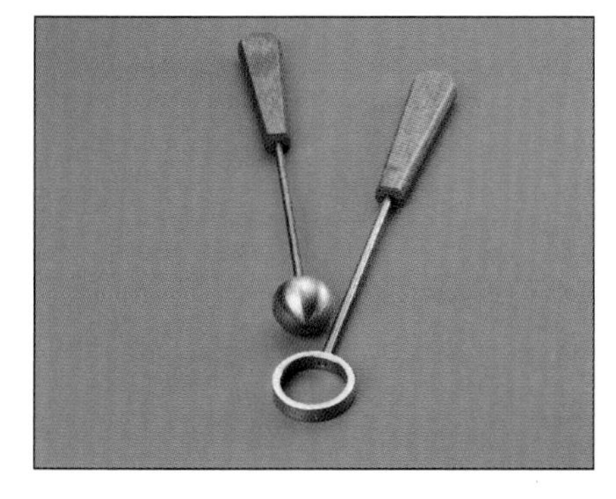

5 Look at the design of the ball and ring apparatus. How will the ball and ring behave when heated or cooled with hot or cold running water? Discuss with your partner(s) what tests you could perform on the apparatus. Perform the tests. (9D)

(a) Describe how you tested the apparatus.

(b) Record what you observed.

Part 3: A Gas

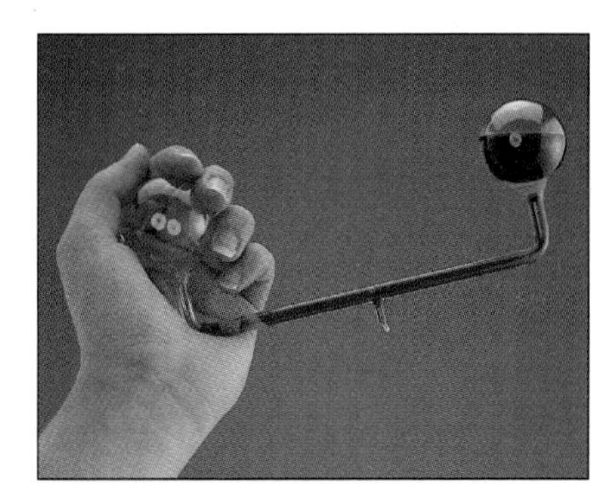

6 Hold one bulb of the pulse glass in your hand and observe what happens. Then hold the same bulb under cold running water and observe again.

(a) Describe what you observed.

Do not heat the pulse glass with the hot plate. Body temperature is enough.

Analysis

7 Analyze your results by answering the following questions.

(a) Did the water expand or contract when it was heated? How do you know?

(b) Predict what would happen to the water level in the tubing if you placed the flask in the refrigerator.

(c) What happened to the solid and gas when they were heated? when they were cooled?

(d) Summarize in a paragraph how the observations you made are significant when designing products.

Measuring Temperatures

Why do you need to measure temperature? Think of all the ways temperature measurement is important in your life. If you know the temperature outside, you know how to dress to go out. If you know the temperature inside your oven, you know if it is hot enough to cook a pizza. No doubt you can think of many more examples.

Many thermometers operate because of expansion and contraction. **Expansion** is an increase in the volume of an object or substance. For most substances, adding heat causes expansion. **Contraction**, which is a decrease in volume, usually occurs when heat is removed from an object or substance. Since solids, liquids, and gases usually expand when heated and contract when cooled, almost any substance can be used in a thermometer.

Not all thermometers rely on expansion and contraction; some rely on electricity or battery power to operate. The choice usually depends on the way the thermometer will be used.

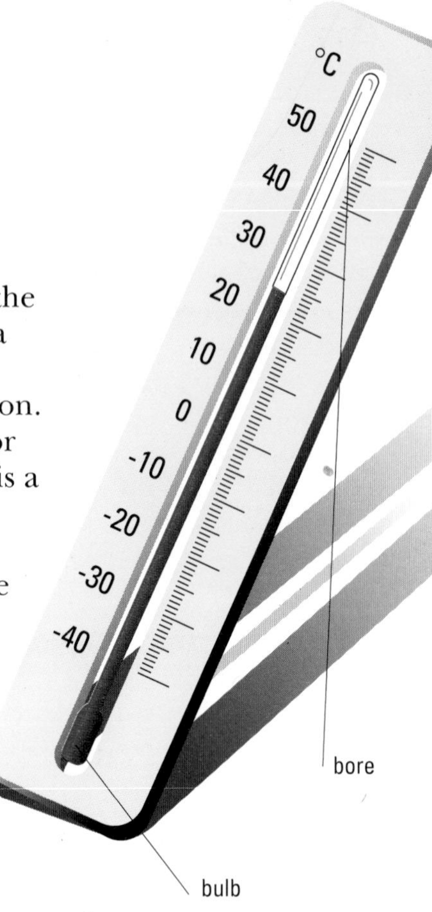

Figure 1
An outdoor thermometer

Liquid Thermometers

You are probably familiar with outdoor thermometers, like the one in **Figure 1**. **Thermometers** use the expansion and contraction of a liquid to measure temperature. The liquid in the bulb (usually coloured ethyl alcohol) expands when it is warmed and is forced up the narrow bore of the thermometer. The more the liquid is warmed, the higher it rises in the bore. When the liquid cools, it contracts, dropping lower in the bore.

The clinical thermometer, **Figure 2**, is a modification of this design. As the liquid is warmed by the patient's body heat, it expands past the constriction to show the patient's temperature. The liquid cools and contracts when the thermometer is removed from the patient, but the liquid cannot move back past the constriction. This allows the patient's temperature to be read long after it is taken. To use the thermometer again, the liquid must first be forced back by shaking the thermometer downward.

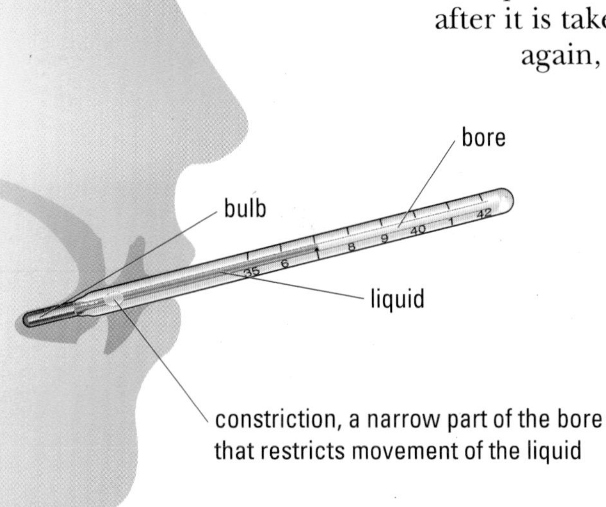

Figure 2
A clinical thermometer. This type of thermometer is sensitive to very small changes in temperature but is able to measure temperatures only within a few degrees of normal body temperature (37°C).

The Thermostat

A thermostat, such as the one in **Figure 3**, can be used to measure temperature in a room or in an appliance, such as a furnace. It can also switch appliances on or off at a preset temperature. In other words, it can act as a feedback system.

Thermostats use the expansion and contraction of solids to measure temperature. They contain a strip made of two metals (called a bimetallic strip). Because the metals are different substances, when they are heated or cooled, the two metals will expand or contract by different amounts, causing the strip to bend. The amount of bending depends on the temperature. This provides a measure of the temperature.

Figure 3

A thermostat can be set to control a furnace to keep the room temperature at 17°C.

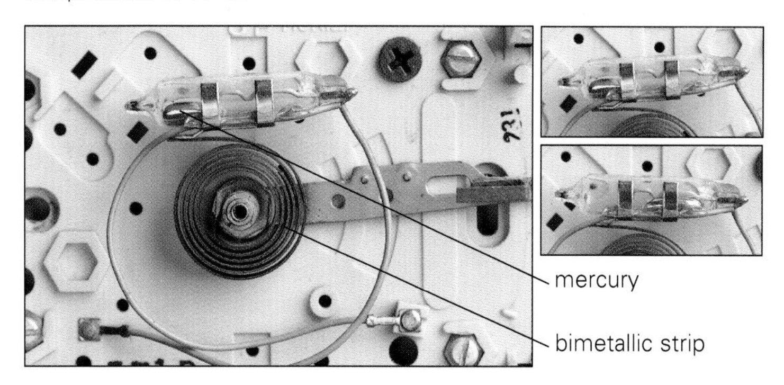

(a) As the temperature in the room rises above 17°C, the bimetallic strip expands. This tilts the glass casing slightly, causing the mercury to flow away from the contact. This breaks the circuit, and the furnace turns off.

(b) When the temperature drops back below 17°C, the strip contracts, causing the mercury to flow to the contact. This completes the circuit, turning the furnace back on.

The Thermocouple

A thermocouple, shown in **Figure 4**, uses electricity to operate. A **thermocouple** contains two wires, each of a different metal, that are joined (coupled) at one end. When two different metals touch each other, a tiny electrical current is generated. The amount of electricity depends on the temperature. The other ends of the wires are connected to a meter that measures electricity. By measuring the amount of electricity that flows through the meter, you can measure the temperature of the metal wires.

Figure 4

The thermocouple is useful for measuring very high temperatures in places where people cannot go, such as inside a kiln or a blast furnace. Also, since the thermocouple creates electricity, it can be connected to a computer to record the temperature.

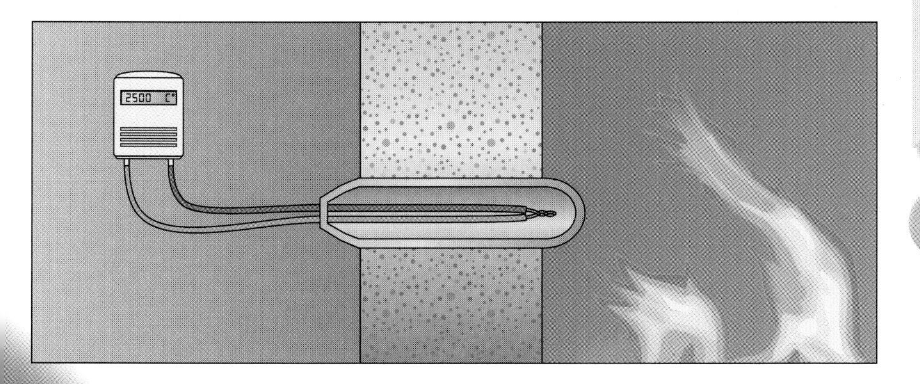

Figure 5

A bimetallic strip

Understanding Concepts

1. Describe in your own words the main features of each device that measures temperature and where it is used.
2. Describe how a thermostat keeps your home at a constant temperature. How does it work as a feedback system?
3. Explain how a thermocouple works. Why can it be connected to a computer?

Exploring

4. The first thermometer was invented by the astronomer Galileo in 1593. Research, using electronic and print sources, and write an explanation describing how it worked.

Design Challenge

Hold a bimetallic strip (**Figure 5**) under hot running water, then under cold running water. Describe what you discover. How can such a strip be used as a switch in a feedback system? Could a bimetallic switch be useful with your Challenge design? Explain.

Temperature and the Tomato

Have you ever wondered why apples grow in the Okanagan Valley and southern Ontario, but not on the Prairies? Or why tomatoes like those in **Figure 1** are often grown in greenhouses rather than in fields? Temperature is an important factor in determining where plants and animals can survive.

The tomato is a native of South America. In the tomato's native land there is no frost, and tomato plants can live for years. In North America, however, tomato plants live and die in less than a year, because they cannot survive if the air temperature drops below 0°C.

Figure 1
Commercial growers of tomatoes often rely on greenhouses to protect their crop from frost.

The Growing Season

Plants begin to grow at temperatures over 5°C. The growing season is the number of days in the year with an average temperature above 5°C. Of course, daily temperatures much higher than 5°C are needed to produce a good crop. **Figure 2** shows the growing season across Canada.

(a) How long is the growing season where you live?

(b) How long is the growing season in Timmins? in St. Thomas?

(c) Which parts of Canada have the longest growing seasons?

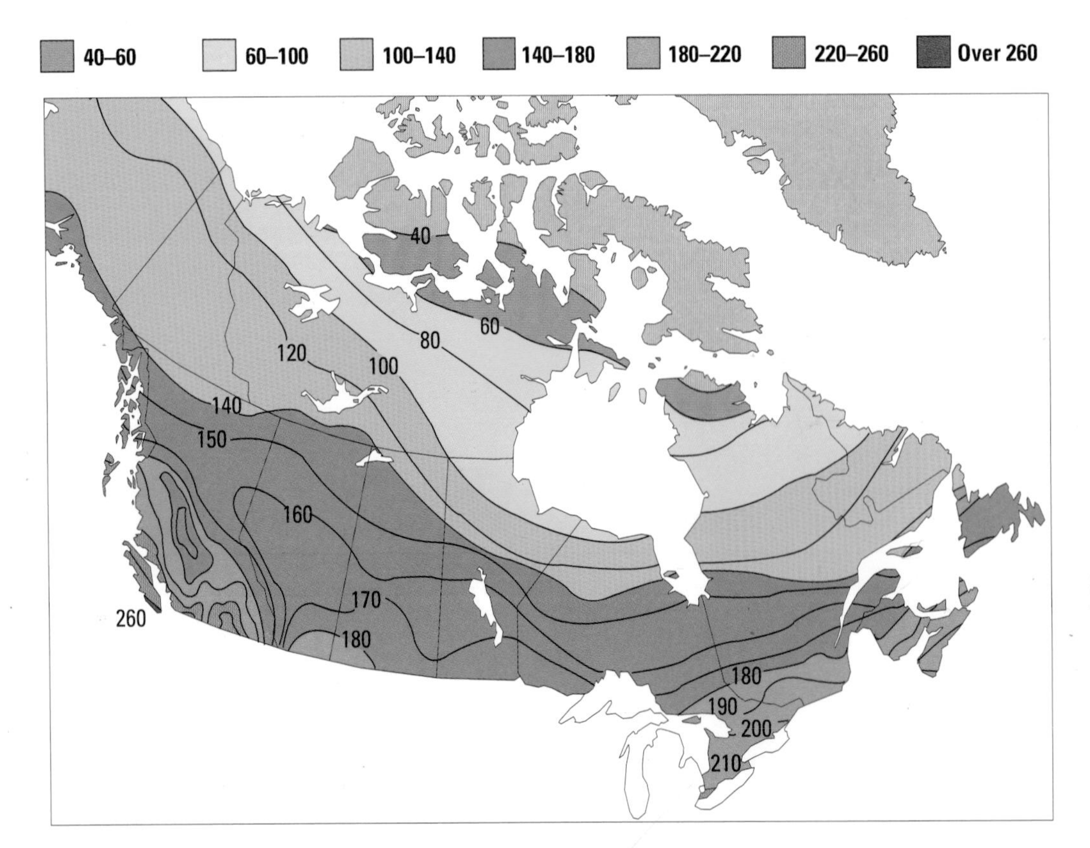

Figure 2
This map shows the number of days in the average growing season. The growing season in any year may be longer or shorter, depending on the weather. Refer to an atlas to locate places.

Variety and the Tomato

The tomato life cycle creates problems for growers in Canada. If you plant a tomato seed outdoors, it will not germinate until the soil is at least 10°C, which may happen 20 or more days into the growing season. After germination the plant will not grow unless night temperatures are higher than 7°C, and it will grow only slowly on cool days.

Tomato breeders have responded by creating varieties that will grow quickly and in cool weather, so they bear fruit in areas with short growing seasons. However, even these varieties, if grown entirely outdoors, will take 60 to 80 d before they grow enough to flower.

After the tomato flowers and is fertilized, the fruit begins to mature. The length of time from flowering until fruit can be picked is called the time to maturity. The amount of fruit that can be picked is called the yield, and is measured in kilograms. The **Tomato Varieties Table** shows the time to maturity for several varieties, and the yield you might expect from each plant.

(d) Using information from **Figure 2** and **Table 1**, which varieties could you grow in your area from seed and expect to get fruit? (Assume 100 d before flowering.) Which varieties of tomatoes could produce fruit in Timmins? in St. Thomas?

(e) Which variety would produce the greatest yield of fruit in your area? in Timmins? in St. Thomas?

Table 1

Variety	Time to maturity (days)	Yield for determinate* varieties (kg per plant)	Yield for indeterminate* varieties (kg per plant every 10 d)
A	80	?	6.8
B	40	?	2.6
C	70	?	5.2
D	60	10	?
E	70	?	4.2
F	65	10	?
G	55	15	?
H	50	10	?
I	75	?	5.6

* NOTE: Some varieties of tomatoes are called determinate—they produce all of their fruit in a short period at the end of the maturity time. Other varieties are called indeterminate—they produce fruit continuously until the first frost. For indeterminate varieties the yield is listed in kilograms per plant every 10 d after maturity. Where in Canada would it be difficult to grow tomato plants, even in a greenhouse? Why?

Helping Mother Nature

Most of Canada has a short period for growing crops. What if you could extend this period? Well, you can! One way is to plant seeds indoors or in a greenhouse. In a warm location, tomato seeds will start to germinate in three to four days. When the plant is about 40 d old, it can be planted in the garden, where it will continue to grow, as long as the soil is warmer than 10°C.

(f) How would starting the seeds indoors change your answers in (e)?

Understanding Concepts

1. Suppose you want to grow as many tomatoes as possible. Plan a growing calendar for a variety listed in **Table 1**. Show planting, germination, flowering, and harvesting dates.

Making Connections

2. In 1998, weather around the world was influenced by El Niño. In much of Canada, temperatures were higher than normal. What effect, if any, might this have on plant growth?

Exploring

3. (4A) You are a reporter for a nature magazine. Choose an animal and research how changes in temperature affect its habitat, its food sources, and its ability to reproduce. Write your information as a magazine article.

Temperature, Heat, and the Particle Theory

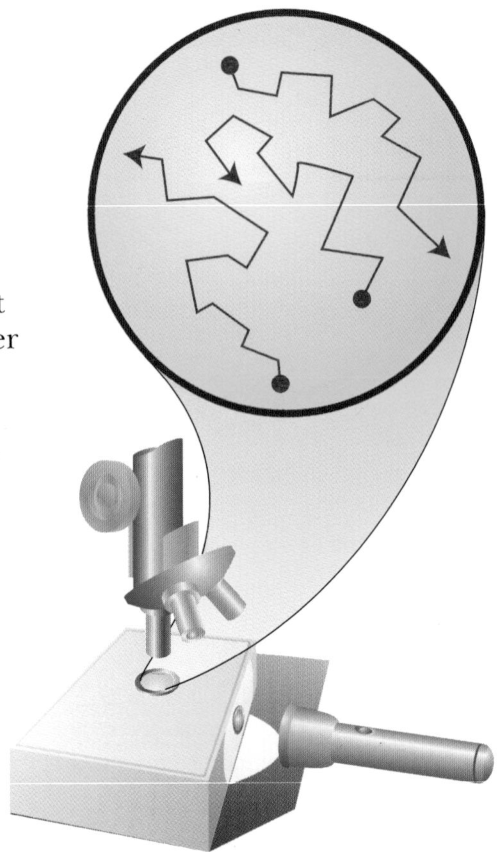

What exactly is the difference between water at 10°C and water at 80°C? What is the difference between heat and temperature? Over the years, scientists have tried to answer these questions with theories about heat and temperature. A theory is an explanation based on all the available information. As new evidence emerges, a theory may change. This is an important part of science.

Many times scientists' theories about big things are changed by observations of very tiny things. For example, in 1827 Robert Brown was looking at pollen. Pollen grains are small—too small to be seen with the eye. Under the microscope they can be seen clearly, if you can get them to sit still. Brown watched pollen grains bounce back and forth, even though the water they were in looked totally motionless. The motion of the pollen grains, now called Brownian motion, provided evidence that helped scientists propose a theory about matter. **Figure 1** shows Brownian motion.

Figure 1
Brownian motion was discovered when pollen grains were observed in water. The same kind of motion can be seen if you look through a microscope at smoke particles.

The Particle Theory

By about 150 years ago, scientists had gathered enough evidence to be able to explain heat and temperature by combining ideas about energy and particles of matter. The kinetic molecular theory, or **particle theory**, says that all matter is made up of tiny particles too small to be seen. According to this theory, these particles are always moving—they have energy. The more energy the particles have, the faster they move.

So far, all the evidence that scientists have about matter supports the idea that all substances are made up of moving particles. That is why we call the particle model for matter a theory. According to this theory, Brown's pollen grains moved because they were being pushed around by invisible particles of water.

a cold water

b hot water

Figure 2
Both hot and cold water are made up of moving particles, some moving quickly, and some moving slowly. Overall, there are more particles moving quickly in hot water than in cold water, so the average energy of the particles is higher.

Heat and Temperature

Heat and temperature are different things. According to the particle theory, **heat** is energy, which transfers from hotter substances to colder ones. **Temperature** is a measure of the average energy of motion of the particles of a substance.

When heat is transferred to particles of cold water, the particles of water move faster, so they have more energy of motion and the temperature of the water rises (**Figure 2**).

Expansion and Contraction

The particle theory is useful to explain why substances expand when they are heated and contract when they are cooled, as you can see in **Figure 3**. At high temperatures, particles have more energy, move more quickly, and have more collisions. As a result, they take up more space, and the substance expands. At lower temperatures, particles have less energy, move more slowly, and have fewer collisions. They take up less space, and the substance contracts.

Figure 3

As a substance is heated, the particles move faster and take up more space.

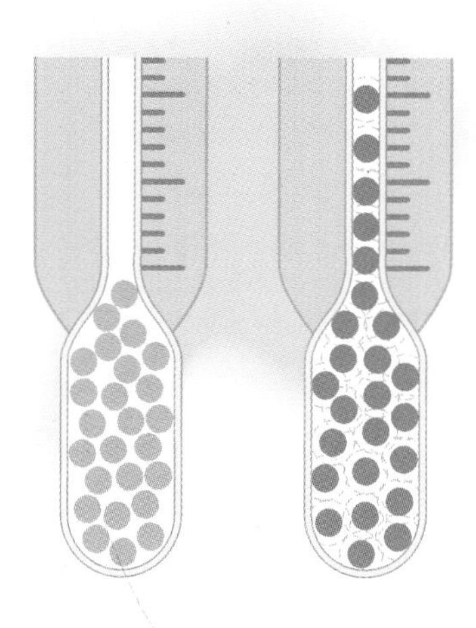

States of Matter

All matter can be grouped into one of three states: solid, liquid, or gas. Each state has certain characteristics, or properties. As shown in **Figure 4**, the particle theory is useful to explain the differences among the properties of solids, liquids, and gases.

Figure 4

The particle theory can be used to explain the properties of the three states of matter.

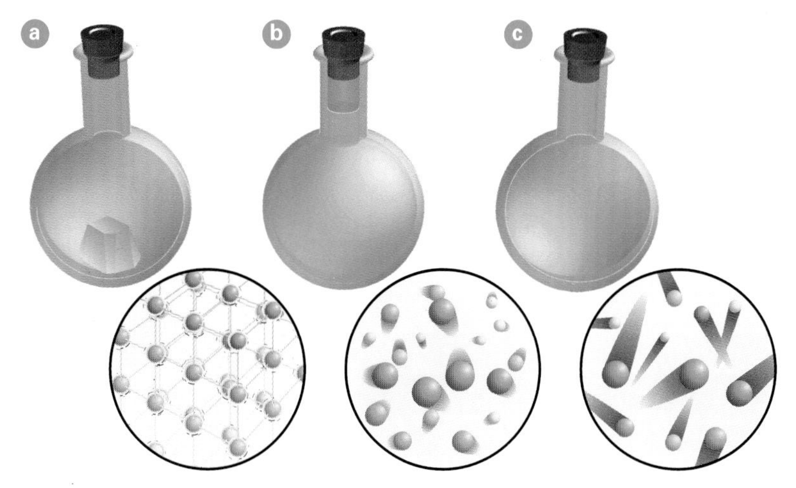

(a) A **solid** has a set volume and a rigid shape; it cannot flow like a gas or a liquid. Particles in a solid move, but only by vibrating in the same spot.

(b) A **liquid** has a set volume, but it will take the same shape as the container it is in; a liquid can flow. Particles in a liquid are free to move around.

(c) A **gas** fills any container it is in and takes on the shape of the container; a gas can flow. Particles in a gas are free to move around and are separated by relatively large spaces.

Understanding Concepts

1. **(a)** What is Brownian motion?
   (b) How did observing this motion help develop a theory about heat and temperature?
2. In your own words, describe the difference between heat and temperature.
3. Explain how a thermometer helps you compare the energies of particles.
4. Use the particle theory to explain how water differs in its three states.
5. Use the particle theory to explain how a clinical thermometer works.

Reflecting

6. How could you use a bottle of perfume in a room to show that the molecules in a gas are in constant motion?

Design Challenge

Is the average energy of particles important in your Challenge? Explain.

SKILLS MENU
- ○ Questioning
- ● Hypothesizing
- ○ Planning
- ● Conducting
- ● Recording
- ● Analyzing
- ○ Communicating

Ice to Water to Steam

Water can exist in the three states shown in **Figure 1**. If you leave an ice cube at room temperature, heat from the surrounding air will transfer to the ice, and the ice will become water. Then if you heat the water enough, it will boil and change into water vapour. In this investigation, you will explore what happens to the temperature of water as it changes state.

Materials

- apron
- safety goggles
- 250 mL crushed ice
- 250-mL Pyrex beaker
- stirring rod
- thermometer
- timing device
- hot plate
- clamp
- stand and ring apparatus
- insulating mitt
- cold metal pie plate

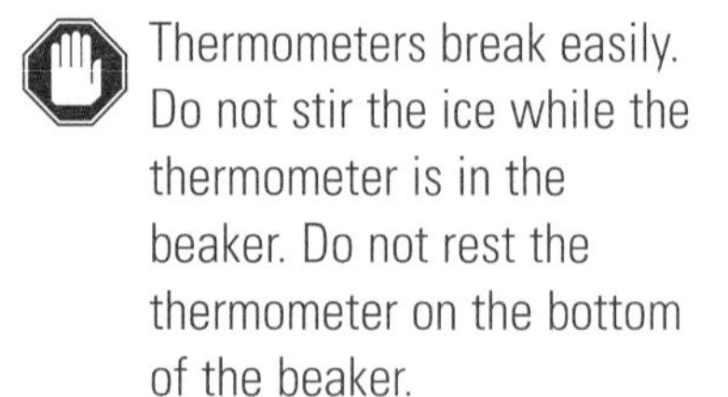
Thermometers break easily. Do not stir the ice while the thermometer is in the beaker. Do not rest the thermometer on the bottom of the beaker.

Question

What happens to the temperature of a substance as it changes state?

Hypothesis

(7C) (2C) **1** Draw a line graph of temperature vs. time, predicting what you think will happen to the temperature of a beaker full of ice as it is heated. Make sure your graph includes temperature values. Write a hypothesis to support your prediction.

Experimental Design

In Part 1 of this investigation you will heat water from ice to a liquid while measuring its temperature. In Part 2 the liquid will be heated until it boils. (For safety reasons, your teacher will demonstrate Part 2.)

(6D) **2** Create a table to record your data.

Procedure

Part 1: Ice to Water

3 Place the crushed ice in a beaker and stir it.
- Place the thermometer in the beaker and record the temperature at 0 min.

(a) Why should you stir the ice/water mixture?

(b) Where should the bulb of the thermometer be to get the most accurate reading? Why?

4 Remove the thermometer and stir the ice again.
- At 1 min, measure the temperature.

(a) Record the temperature each minute until 5 min after all the ice has melted.

(b) Record your observations of any changes.

Part 2: Water to Steam

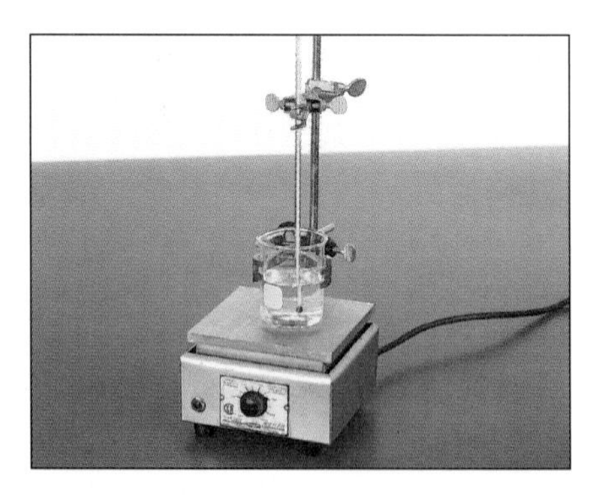

5 To find out what happens to the temperature of water as it boils, a demonstration will be set up as shown above.

(a) Record temperature observations every minute before the water boils, and for at least 5 min after it begins to boil.

SKILLS HANDBOOK: (7C) Constructing Graphs (2C) Predicting and Hypothesizing (6D) Creating Data Tables

Figure 1
Water can be solid (ice), liquid, or gas (water vapour).

Making Connections

1. While camping in the fall, you leave some water in a bucket overnight. The next morning, you notice a layer of ice on the top of the water. What is the temperature of the water just beneath the ice?

Exploring

2. Design an experiment to find out if the melting temperature (2E) of water changes for different amounts of water. Check your design with your teacher before starting.

Reflecting

3. Compare the graph you drew for your hypothesis with the graphs you drew from your data. Was your hypothesis correct? Explain.

Part 3: Steam to Water

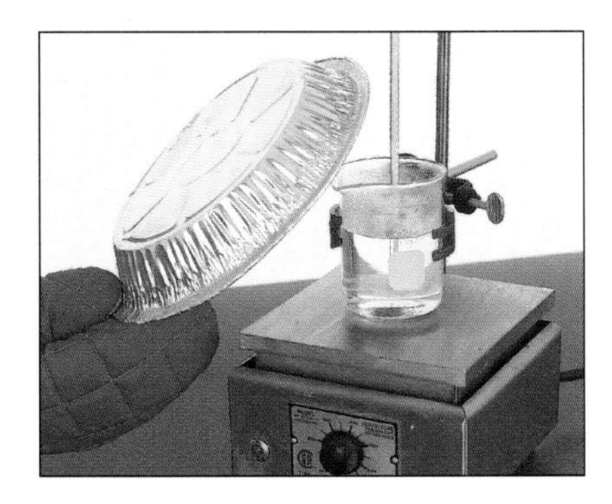

6 Use an insulating mitt to hold a cold metal pie plate above the boiling water for 10 s.

(a) Describe what you observe.

Analysis

7 To analyze your results, answer the following questions.

(a) Create line graphs of temperature vs. time (7C) for your data from Parts 1 and 2.

(b) Describe the shape of (7B) the graphs.

(c) Predict what your graph would look like if you had continued to heat the melted ice.

(d) Predict what your graph would look like if you had continued to heat the steam.

(e) What happens to the temperature of water during a change of state?

(f) On the basis of your observations, do you agree with the following statement? "When heat is added to a solid, the heat can cause a change of state or an increase in temperature." Explain your answer.

Design Challenge

Describe how you might use tables, charts, or graphs to present the data from your Challenge.

(7B) Reading a Graph (2E) Designing an Inquiry Investigation

The Particle Theory and Changes of State

You found that adding heat to ice at 0°C did not cause much change in temperature—it simply changed the ice to water. The temperature at which a substance melts is called its **melting point**. For ice, that point is 0°C.

Adding heat to water at 100°C does not cause a change of temperature—it simply changes the water to steam or vapour. The temperature at which a substance boils is called its **boiling point**. What is the boiling point of water?

In **Figure 1**, the particle theory of matter is used to explain why the temperature of a substance remains constant during a change of state.

Figure 1

In (a) water is changing from ice to water. In (b) water is changing from a liquid to water vapour.

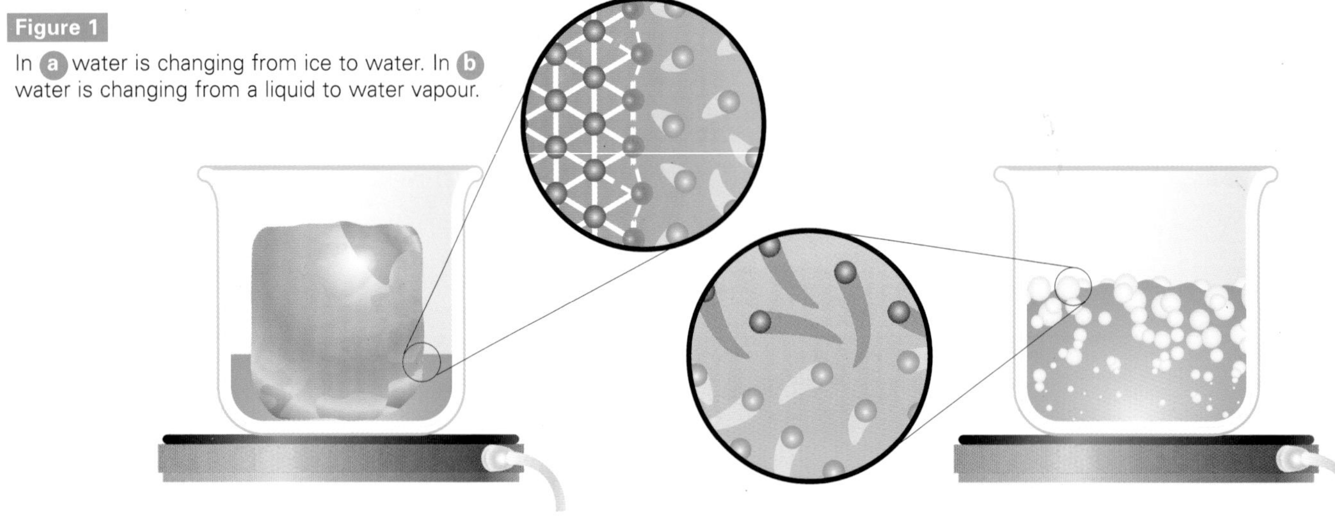

(a) When you add heat to ice at 0°C, the temperature does not rise. Instead, the heat is used to free the water particles from their set places in the solid. Since the heat has not increased the motion of the water particles, their temperature is the same. Particles that have been set free are in the liquid state.

(b) As water is heated, its particles gain energy and now move faster and faster. The temperature rises until the liquid reaches its boiling point. At this point the temperature does not continue to rise. Instead, the heat is used to help the particles move more freely and escape from the surface of the liquid. Particles that have been set free are now in the gas state.

Heating Curves

A **heating curve** is a graph showing how a substance's temperature changes while being heated. An example is shown in **Figure 2**. The flat part of the curve, where the substance is melting or boiling, is called a plateau. What do you think a cooling curve is?

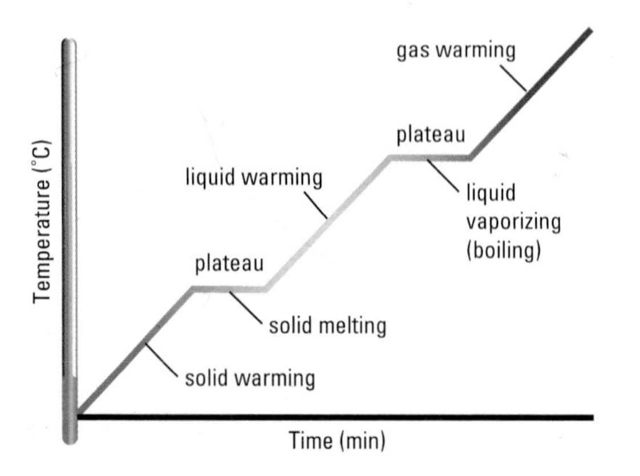

Figure 2

A heating curve

Differences in Substances

Different substances are made of different particles, so they have different melting and boiling temperatures.

Substance	Melting point/ freezing point (°C)	Boiling point/ condensation point (°C)	Some uses for substance
oxygen	–218	–183	• welding • breathing apparatus
carbon dioxide	–78	–78	• as "dry ice" in refrigeration
ammonia	–78	–33	• fertilizer • explosives
mercury	–39	357	• barometers • switches
ethyl alcohol	–114	78	• solvent
water	0	100	• ice cubes • drinks
paraffin wax	71	360	• candles • sealing jars of jam and jelly
aluminum	660	2467	• auto and aircraft parts • window and door frames
iron	1535	2750	• cast iron • wrought iron • steel
tungsten	3410	5660	• light bulbs • cutting tools

Changes of State

Figure 3 sums up changes in state. Notice that there are two changes of state called sublimation. The liquid state is skipped when frost forms from water vapour on windows in winter, or when a solid air freshener slowly disappears.

Figure 3
When heat is absorbed or released by a substance, a change of state can occur. Heat that is absorbed causes the changes shown by red arrows. Heat that is released causes the changes shown by blue arrows.

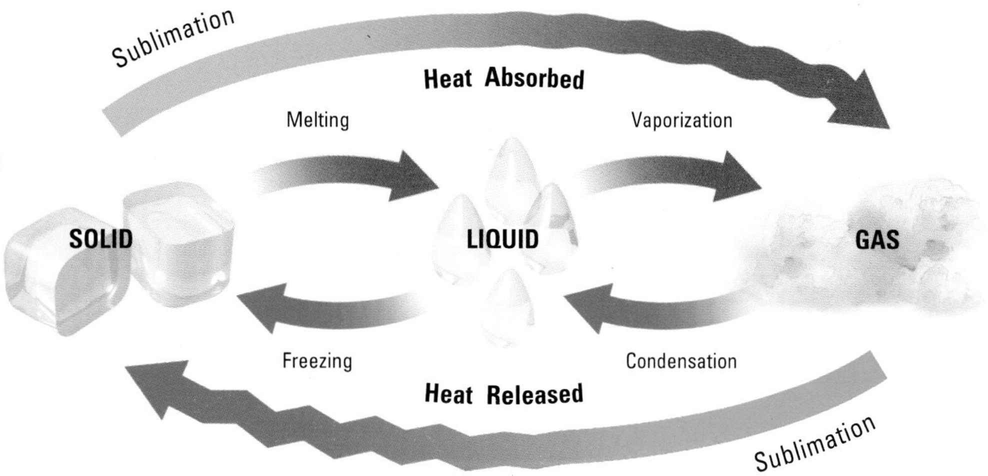

Understanding Concepts

1. Use the particle theory and a series of diagrams to explain what happens to ice as it is gradually heated and turns to water and then steam.
2. Draw a heating curve for paraffin wax as it goes from 0°C to 400°C. Label its melting point, boiling point, and its state in each section of the curve.
3. Draw a cooling curve for steam to (7C) water to ice. Label your graph.

Making Connections

4. Look at the melting and boiling points of mercury and ethyl alcohol. Which of these two substances would be better to use in an outdoor thermometer in the Arctic? Why?
5. Sweating helps cool a person's body.
   - **(a)** What change of state does sweating involve?
   - **(b)** Why does sweating cause cooling?

Exploring

6. The Celsius temperature scale relies on fixed points, the melting and boiling points of water, and divides the interval between them into 100 degrees. The reason water was chosen is that it is a common and very useful substance. Make up a temperature scale for a planet where the oceans and lakes are filled with ammonia instead of water.

SKILLS HANDBOOK: (7C) Constructing Graphs

2.8 Inquiry Investigation

SKILLS MENU
- ○ Questioning
- ● Hypothesizing
- ○ Planning
- ● Conducting
- ● Recording
- ● Analyzing
- ● Communicating

Heat and Convection

Do all parts of your classroom feel as if they have the same temperature? Does the air near the floor or window feel the same as the air near the middle of the room? Heat can cause the air in a room to move. In this investigation, you will learn how heat transfers (or moves) in air and other fluids (**Figure 1**).

A fluid is a substance that is free to flow. Liquids and gases are fluids. To transfer heat, fluids use a method called convection. **Convection** is the transfer of heat by the movement of particles from one part of a fluid to another. The motion of many fluid particles is called a **convection current**.

Question

What are the properties of a convection current?

Hypothesis

1 Read the procedure for this experiment. Then draw diagrams predicting how you think the pieces of nutmeg and the smoke will move. Write a hypothesis for your prediction. (2C)

Experimental Design

In this experiment you will use markers (the nutmeg in Part 1 and the smoke in Part 2) to reveal the movement of two fluids.

Materials

- apron
- safety goggles
- grated nutmeg
- large Pyrex beaker
- cold water
- retort stand and clamp
- hot plate
- ring stand
- gas convection apparatus
- candle
- wire gauze
- smoke paper

Follow the safety rules in the *Skills Handbook* when using an open flame.

Procedure Part 1: Convection in a Liquid

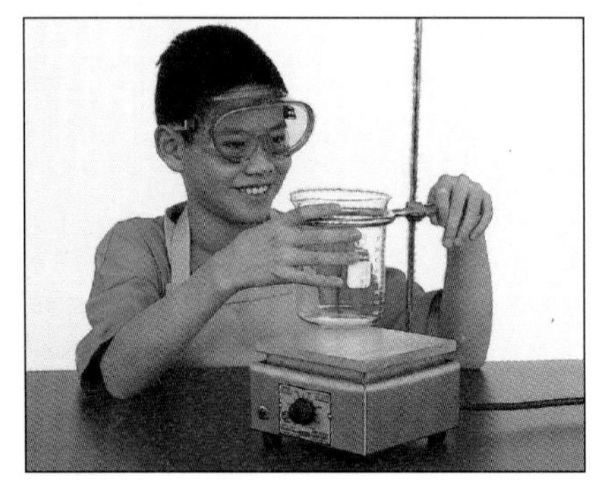

2 Set up the beaker so that one side will be heated more than the other.

3 Fill the beaker with cold water. Sprinkle about 15 to 20 pieces of nutmeg into the water.

(a) What do you think will happen to the pieces of nutmeg? Why?

4 When the water and pieces of nutmeg have become still, turn on the hot plate.

5 Heat the water until you can see a pattern in the motion of the nutmeg pieces.

(a) Record your observations.

(b) Draw a diagram showing the pattern you observed.

(c) Beside your diagram, describe any evidence that there is a convection current.

Figure 1

"Heat waves" are caused by rising air.

Part 2: Convection in a Gas

6 Your teacher will set up the gas convection apparatus with a candle on the right side.

(a) Predict what will happen when smoke paper is held first above the right chimney, then above the left chimney. Record your prediction.

(b) Record your observations.

(c) Draw a diagram showing the pattern you observed.

Analysis

7 Analyze your results by answering the following questions.

(a) You set up your experiment in Part 1 to cause uneven heating. In your drawing from step 5, show where you think the water would be warmer, and where it would be cooler.

(b) You have learned that substances expand when they are heated, because they are made of particles. Using the particle theory, explain how a convection current could start in a fluid.

Making Connections

1. (a) Predict how convection currents will form in the air in your bedroom when the furnace or heater is turned on. Sketch your prediction.

(b) How could you investigate your prediction?

2. (a) Your town council decides to heat the town's swimming pool. If you had to put a heater directly in the pool, where could you install it to take advantage of convection currents? Explain.

(b) The town council decides to put the heater and a pump in a shed away from the pool. Design a circulating system for the heater that takes advantage of convection currents. (8B)

(c) Explain why the town council decided not to put the heater in the pool.

Design Challenge

How can you adapt your design for the town pool in question 2(b) to help you with your design for a solar-heated pool?

SKILLS HANDBOOK: (2C) Predicting and Hypothesizing (8B) Creating a Design Folder

Heat and Weather Patterns

Heat transfers in fluids using convection currents. Convection currents in Earth's atmosphere and oceans are important because they affect weather everywhere.

Thermals, Winds, and Sea Breezes

On a clear day, energy from the Sun warms the land and the air near it. The warm air expands and becomes less dense. (Can you use the particle theory to explain why this happens?) As the warm air expands it rises and is replaced by cooler, denser air. This kind of convection current is called a thermal. **Figure 1** shows the formation of a thermal.

As the thermals grow in size, drawing more and more cool air in at the bottom, people on the ground can begin to detect the movement of air. We call the flow of air a wind. This kind of wind is most likely to develop near a large body of water, such as a large lake or an ocean. Water warms up more slowly than soil, so the air over the water tends to be cooler. The cooler air over the water moves toward the base of the thermals on land. If the convection current is near a sea or an ocean, the wind that forms is often called a sea breeze.

Thermals help large birds, such as turkey vultures and eagles, glide in the air for hours, searching for prey. As the warm air rises, the birds can ride it higher by gliding in circles within the thermal. When the thermal isn't strong enough to keep them aloft without effort, or they want to move to a new area, they glide down to the next thermal and start the cycle again.

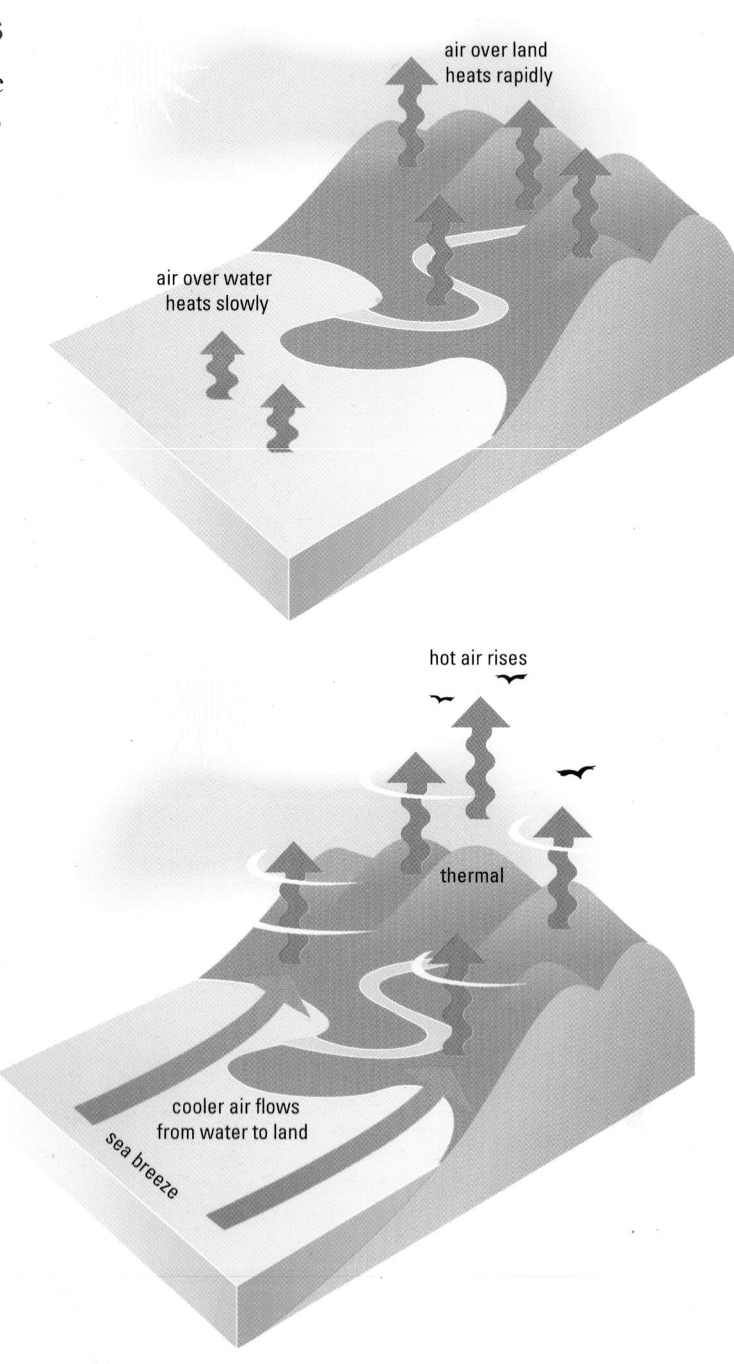

Figure 1
Heating of the air by the Sun causes thermals and sea breezes to form.

SKILLS HANDBOOK: 4A Research Skills 8C Multimedia Presentations

Large Wind Patterns

What we see in thermals is also happening on a much larger scale. As you can see in **Figure 2**, huge convection currents are created by the Sun in Earth's atmosphere, causing global wind patterns to develop.

Based on the simple convection model, you might expect winds at Earth's surface to blow only from the poles to the equator. But the circulation of air is much more complicated than that. One factor that affects wind direction is Earth's motion. Earth spins on its axis once each day. As it spins, the air that is rising and falling can also start to move in curved or circular patterns. (This effect is like water going in circles as it empties from a sink.) The result is that air moves in general patterns, but at a local level wind direction and strength are difficult to predict—as you can tell from faulty weather forecasts! Global wind patterns are shown in **Figure 3**.

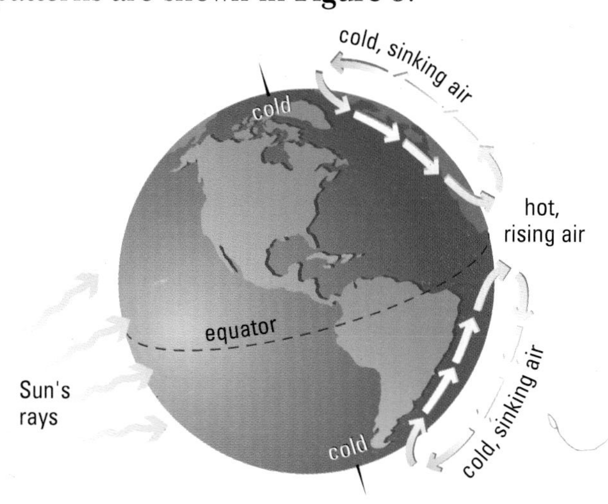

Figure 2

Air near the equator is much warmer than air near the poles. The warm, less dense air rises above the equator. It starts to move north and south, away from the equator, as cooler, more dense air moves in to replace it. Large convection currents of air are set up when hot air near the equator rises.

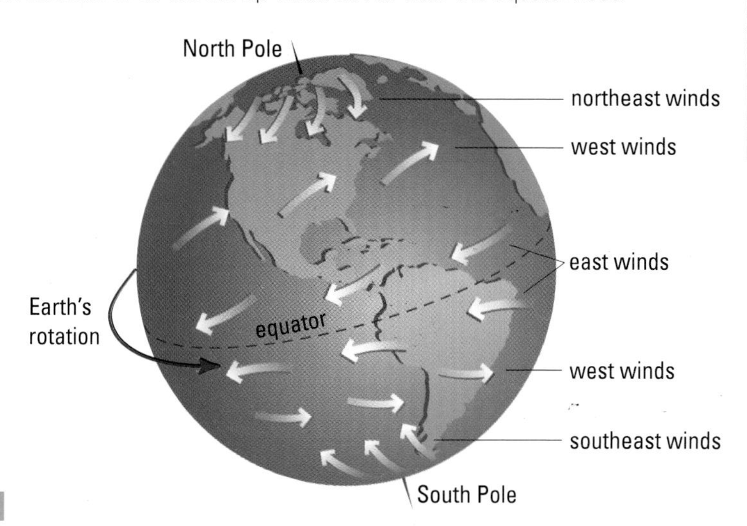

Figure 3

Earth's motion causes patterns in the large movements of air in the atmosphere. Notice that the winds near the equator tend to travel westward. The winds in southern Canada tend to travel from west to east. What direction would you expect the wind to blow on most days in Dawson City?

Understanding Concepts

1. Explain the expression "hot air rises."
2. (a) Are thermals more likely to be set up in winter or in summer? Why?
   (b) Are thermals more likely to be set up on a cloudy day or a sunny day? Why?
3. Draw and label a sequence of three diagrams to show how a thermal is set up over land.
4. What conditions help a sea breeze begin?
5. A land breeze, blowing from the land out over the water, can begin when the sun sets. When this happens, lakes often become calm near the shore. Use a diagram to help you explain how a land breeze is set up.
6. Describe the main factors that cause the air in Earth's atmosphere to move.

Exploring

7. (4A) (8C) Using electronic and print sources, research the sport of hang-gliding. Relate what you discover to thermals and the way birds soar on thermals. Make a visual presentation about what you discover.
8. (4A) Hurricanes are large storms that occur on and near the Atlantic Ocean. Research how these storms are set up by heat transfers.

Design Challenge

You have learned that heat transfer in water and air sets up convection currents. How can knowledge of convection currents help you with your challenge design?

2.10 Design Investigation

SKILLS MENU
- ○ Identify a Problem
- ○ Planning
- ○ Building
- ● Testing
- ● Recording
- ● Evaluating
- ● Communicating

Heat and Conduction

Imagine that your first few steps in bare feet in the morning are on a rug. Then you walk on to a wooden floor. Which do you think would feel cooler? Can you suggest a reason?

The wooden floor feels cooler than the rug even though they are at the same temperature. Wood has a better ability than a rug to transfer heat away from your body. You can notice a similar effect when you touch objects around you in the room: a metal chair leg transfers heat away from your fingers more quickly than a piece of paper.

The transfer of heat by the collisions of particles in a solid is called **conduction**. **Figure 1** shows conduction. A substance that conducts heat well is called a **heat conductor**. Wood is a better heat conductor than a rug, and metal is a better heat conductor than paper. In this investigation you will test substances to rank them according to which are the best heat conductors.

Materials

- apron
- safety goggles
- glass and metal rods of approximately equal size
- 2 support stands
- 2 clamps
- candle
- hot plate
- equal-sized rods made of various metals
- timing device

Use care with an open flame.
Do not touch wax while it is liquid. Hot wax can burn.

Do not allow the glass or metal rod to touch the hot plate.

Problem

You are going on a camping trip and will be cooking over an open fire.

Design Brief

Test a variety of materials to determine which would be the most suitable cookware to design and make for a camping trip.

Design Criteria

- A utensil is necessary for cooking hot dogs and making hot chocolate.
- The material should conduct heat quickly and be safe to hold.

Test

Part 1: Comparing a Metal and a Nonmetal

1 Clamp a glass rod and an equal-sized metal rod to separate support stands.
- Use a lit candle to make small beads of wax equally spaced along the rods.

2 After the wax beads are solid, arrange the rods horizontally so that each rod has one end above the hot plate.
- Turn on the hot plate and time how long it takes each wax bead to melt.

(a) Record your observations.
- Turn off the hot plate after the last bead has melted.

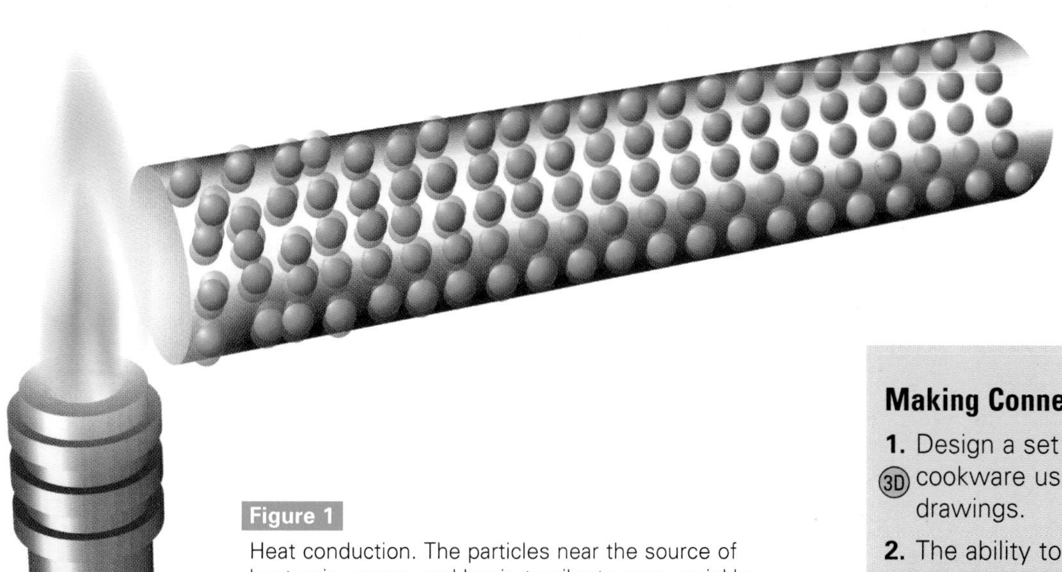

Figure 1
Heat conduction. The particles near the source of heat gain energy and begin to vibrate more quickly. They collide with other particles and transfer their energy to them.

Part 2: Comparing Metals

3 Design a way to test various other metal rods so you can rank the metals from best conductor to worst.

- Design a table for your data.
- Have your teacher approve your design.
- Carry out your test.

Evaluate

4 Evaluate your results by answering the following questions.

(a) In step 2, did the wax melt more quickly on one rod than the other? Explain why.

(b) When conducting this test, what were the independent and dependent variables? List the controls that you used.

(c) Which materials would be best suited for camping cookware?

Making Connections

1. Design a set of camp (3D) cookware using technical drawings.
2. The ability to transfer heat quickly is important in products we use in our homes. Describe three products that transfer heat through a solid by conduction.
3. A baker can choose between using a glass pan or a metal pan. In which pan would bread develop a firm brown crust on the bottom and sides? Explain how.
4. Explain why cooks stick metal objects in potatoes when they bake them.

Exploring

5. Describe how you could use a set of temperature probes connected to a computer to test how quickly various solids transfer heat.

Reflecting

6. Use the particle theory to explain how heat is transferred along a metal rod.

Design Challenge

Substances can be good or poor conductors of heat. How will this affect your choice of materials for your Challenge? How could you test your materials?

SKILLS HANDBOOK: (3D) Planning a Prototype

2.11 Career Profile

At the Scene of the Crime

Susan Kern is a forensic biologist at Ontario's Centre of Forensic Sciences. That means Susan spends a lot of time looking at blood and other body fluids. "We determine if a particular body fluid is present on an item," she says. "If it is, we try to determine who the fluid may have come from, using DNA typing."

Susan sometimes goes to the crime scene to help figure out what happened. "The size, shape, and pattern of blood stains can tell us a lot about what occurred," she says. A quick chemical test tells Susan whether a stain is probably blood, and she uses a small hand-held magnifier to measure very small stains.

After she has analyzed all the evidence and made her conclusions, she goes to court as an expert witness. "Being in the witness stand can be stressful, but it's the culmination of all we do," she says.

Susan enjoys her work and finds it challenging, but what about the gory crime scenes, do they upset her? "Not really—it takes a certain type of person to look at it as just work. It's an interesting job."

Heat and Crime

Often, the temperature of an object plays an important role in solving a crime. Heat from the engine of a vehicle may indicate that it has just been driven. A warm cup of coffee on a kitchen table may mean that a suspect has just left. The temperature of a body can indicate when death occurred.

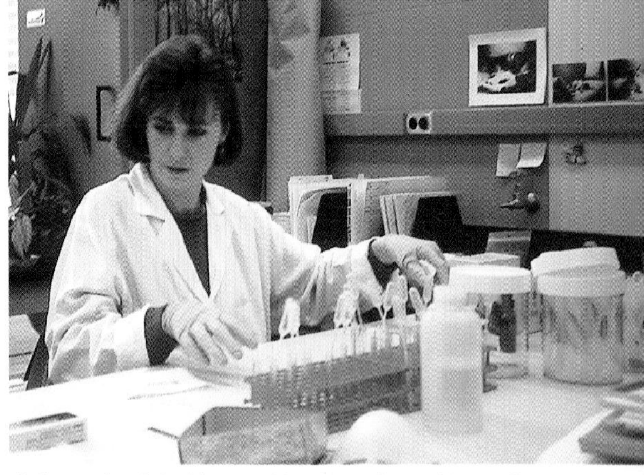

A forensics lab. Here evidence collected by police officers is analyzed and tested.

When an object is removed from a source of heat, its temperature begins to drop. It cools until it reaches the temperature of the surrounding air, at which point its temperature stabilizes. For same-sized objects at the same initial temperature, surrounded by material with the same insulating properties, heat loss occurs at the same rate. This allows forensic scientists to determine, with some degree of accuracy, the time that the object was removed from the heat source.

At death, a body stops generating heat and its temperature drops gradually. At first the cooling is relatively rapid—approximately 0.8°C per h—but this rate slows after a few hours. In addition to the initial temperature, the temperature of the surroundings, and the effectiveness of any insulation, the size of the body will also affect the cooling time. A small body will cool more rapidly than a large one. By taking a body's temperature, a forensic scientist can estimate the time of death.

The Alibi—A Forensic Investigation (2A)

A suspect in a jewel robbery tells police that he was at home pouring a cup of coffee when the theft occurred. The theft took place at 10:00 p.m. When police arrived at 10:15 p.m., the temperature of the suspect's coffee was 45°C. Was he telling the truth?

- Make a data table like the one below.

Time (min)	Temperature (°C)
0	?
1	?
2	?
3	?

- Pour 250 mL of hot coffee into a coffee cup.
- Put a thermometer into the coffee and measure the temperature of the coffee.
- Record the temperature at 0 min.
- Measure and record the temperature each minute for 20 min.
- Make a graph of your data.

1. Does your data support the suspect's alibi?
2. How could your graph help in other investigations?
3. How would your data be affected by changing the insulating properties of the cup?

2.12 Inquiry Investigation

SKILLS MENU
- ○ Questioning
- ● Hypothesizing
- ● Planning
- ● Conducting
- ● Recording
- ● Analyzing
- ● Communicating

Radiation

You have learned that heat transfer occurs by convection in fluids and conduction in solids. Both convection and conduction depend on the motion of particles. Between Earth and the Sun there is much empty space—there are almost no particles. Yet we receive heat from the Sun; if we didn't, there would be no life on Earth. There must be a third way of transferring energy, one that does not need particles. This method is called radiation. **Radiation** is the transfer of energy by means of waves. Energy transferred by radiation is called **radiant energy**. Light is an example of radiant energy.

Some objects absorb radiant energy better than others; some objects emit (give off) radiant energy better than others. In this investigation you will test a variable to see what effect it has on absorbing or emitting radiant energy.

Questions

(a) How can the absorption of radiant energy be increased?

(b) How can the emission of radiant energy be increased?

Hypothesis

1 (2C) Your experiment will study one of the two questions, and one independent variable. For your experiment, create a hypothesis in the form "If we…, then…."

Experimental Design

2 You will create a controlled experiment that studies the effect of changing either the colour or the surface area of an object. **Figure 1** may help you with your design.
- You should use water as the substance to be heated or cooled. Any other materials (3D) you choose should be inexpensive and safe to use.
- The experiment must be controlled. While you are carrying out your tests, you must change only one variable at a time.
- Since different groups will be testing (2D) different variables, you must be prepared to share your results.

3 Discuss with your partner(s):
- the experimental design
- the materials you will need to test your hypothesis
- how you are going to record and present the data you collect

4 Prepare a procedure, including safety precautions.

Materials

5 List the materials you will need.

6 Ask your teacher to approve your procedure and materials list.

Procedure

7 Carry out your experiment.

8 Share your results with the other groups in your class.

SKILLS HANDBOOK: (2C) Predicting and Hypothesizing (3D) Planning a Prototype (2D) Identifying Variables and Controls

Figure 1

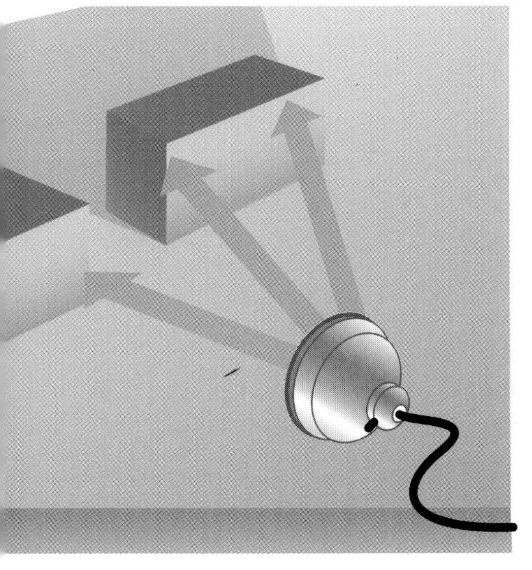

(a) Which object will absorb radiant energy fastest? Why?

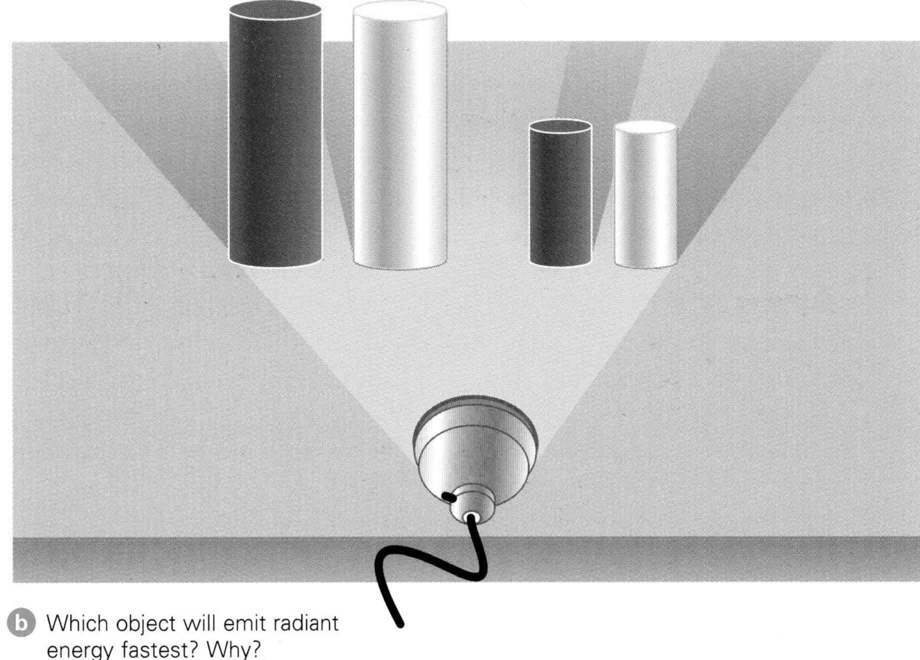

(b) Which object will emit radiant energy fastest? Why?

Analysis

9 Analyze your results by answering the following questions.

(a) Which colour or colours tend to absorb radiant energy best?

(b) How does the surface area of an object affect its ability to absorb radiant energy?

(c) Which colour or colours tended to emit radiant energy best?

(d) How does the surface area of an object affect its ability to emit radiant energy?

(e) Does an object that absorbs radiant energy well also emit radiant energy well? Explain your answer.

(f) In your experiment, what was the independent variable? What was the dependent variable? What variables did you control?

(g) If you were to try this experiment again, what would you do to improve your choice of materials and procedure? Explain.

Exploring

1. (2C) (2E) Are good conductors or poor conductors better emitters of energy? Describe how you would test your hypothesis. If possible, carry out the experiment and share the results with your class.

Reflecting

2. What are some properties of materials that get hot when solar energy hits them? Explain.

Design Challenge

Should the materials you choose for your challenge design be good absorbers of radiant energy or good emitters? What changes should you make in your design based on your knowledge of radiant energy?

Heat and the Water Cycle

Without water, life as we know it would not be possible. The water cycle, **Figure 1**, is the repeated movement of water from oceans and lakes, to atmosphere, to land, and back to oceans and lakes. The cycle begins when radiant energy from the Sun is absorbed by the oceans and other bodies of water.

1. Raindrops

High above Earth's surface, the temperature is lower, so the water vapour cools. Tiny water droplets form around some of the dust particles in the air. The droplets or crystals keep growing until they are large enough to be pulled down by gravity. As they fall, they meet other droplets and form drops of rain.

2. Freezing Rain

Freezing rain forms and falls like ordinary rain, but freezes when it hits cold surfaces. **Figure 1** shows the conditions needed to cause freezing rain. Usually these conditions do not last very long, but in one case, the 1998 ice storm in Quebec and eastern Ontario, they lasted several days.

Figure 1

Earth's water cycle starts with water vapour from the surface of Earth, including oceans, lakes, and living things, rising with air currents.

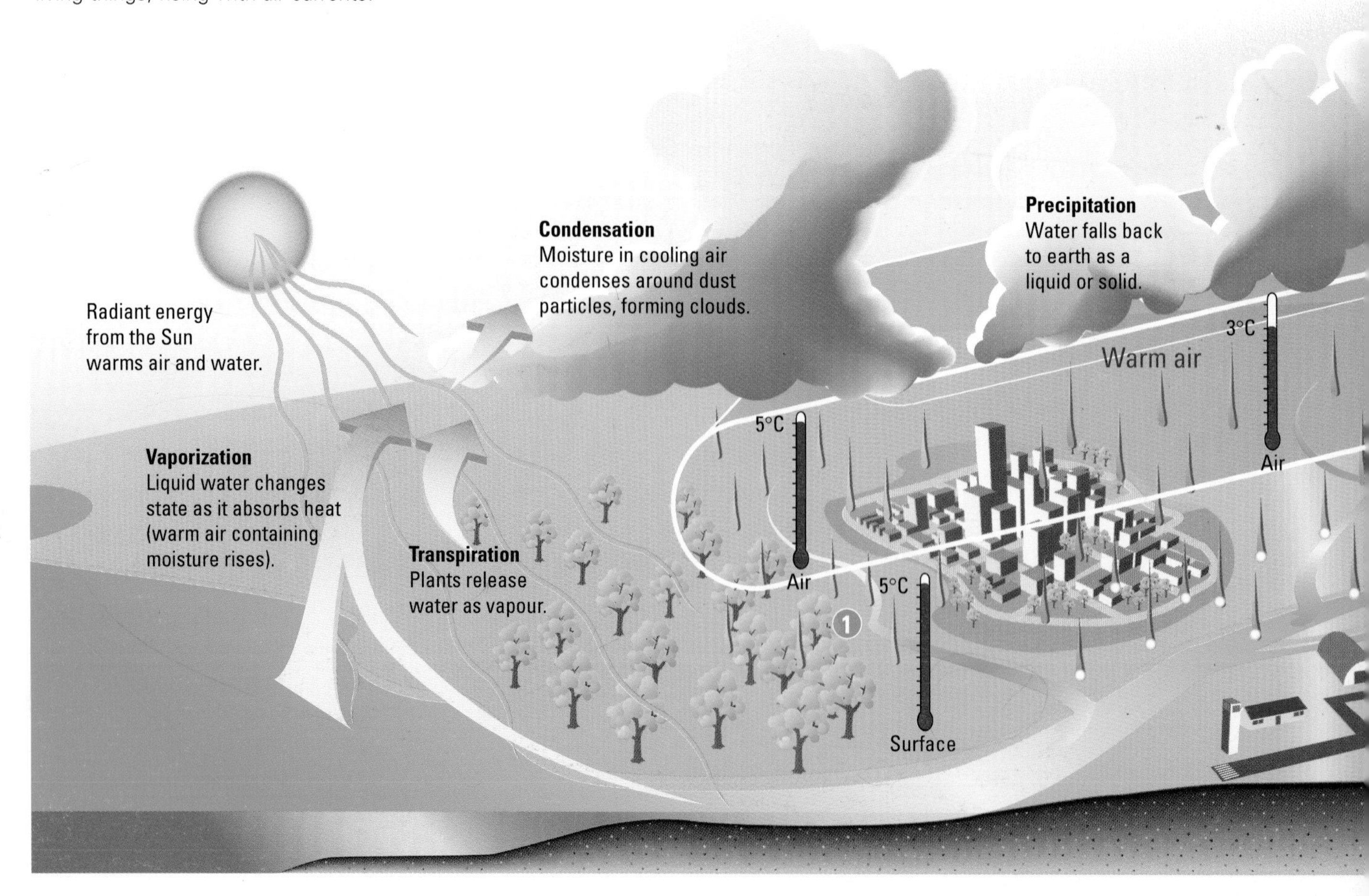

Weather Response

Freezing rain can build upon power lines and pull them down, cutting off electricity. It can make roads and sidewalks slippery and dangerous to use. A weather advisory predicts freezing rain for your area.

(3D) 1. Design a device to reduce the effects of freezing rain.

3. Snow

At colder temperatures, snowflakes form instead of raindrops.

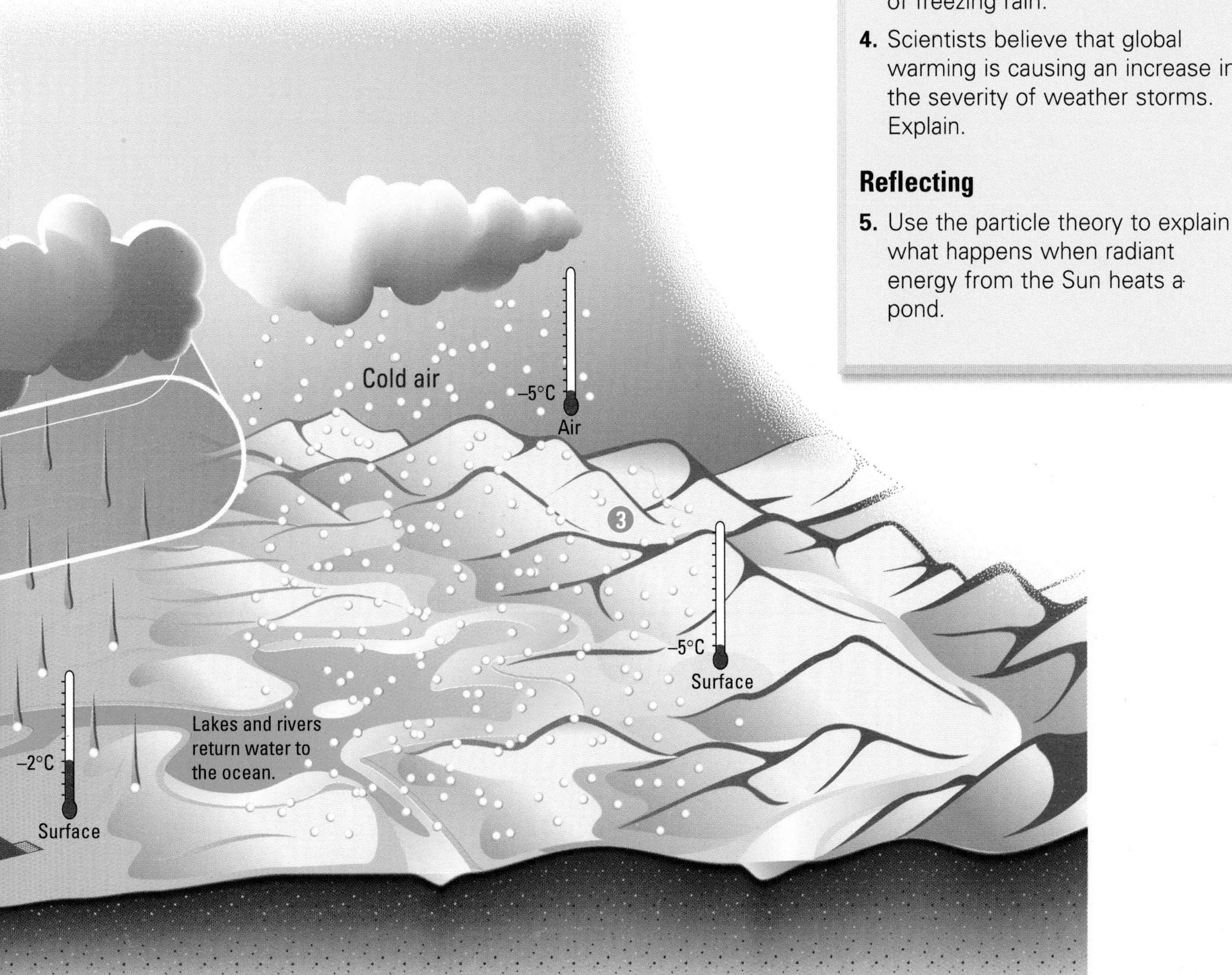

Understanding Concepts

1. What is the main source of energy for the water cycle on Earth?
2. (a) Starting with water on the Earth's surface, list all the changes of state that occur in the completion of one water cycle when rain falls.
   (b) How does the list of changes of state differ for a snowfall?
   (c) How does the list differ for freezing rain?

Exploring

3. Hail, like rain and snow, is a type (4A) of precipitation. It can damage property and destroy farmers' crops. Research how hail is formed. Draw a diagram showing how its formation differs from that of freezing rain.
4. Scientists believe that global warming is causing an increase in the severity of weather storms. Explain.

Reflecting

5. Use the particle theory to explain what happens when radiant energy from the Sun heats a pond.

Heating Homes

Think about heating the place where you live: Where does the energy come from to heat your house or apartment? How does the heat get sent to each room? Which of the three methods of heat transfer (convection, conduction, and radiation) occur? Does the heating system use energy efficiently? Is there a feedback device to control the system? Try to relate these questions to the ideas about home heating described here.

Heating a Single Room

Air is a good heat insulator, which, of course, means it is a poor heat conductor. So in order to efficiently heat air in a room, a convection current must be set up. **Figure 1** shows how a convection current is created by an electric heater.

Figure 1

An electric heater creates a convection current.

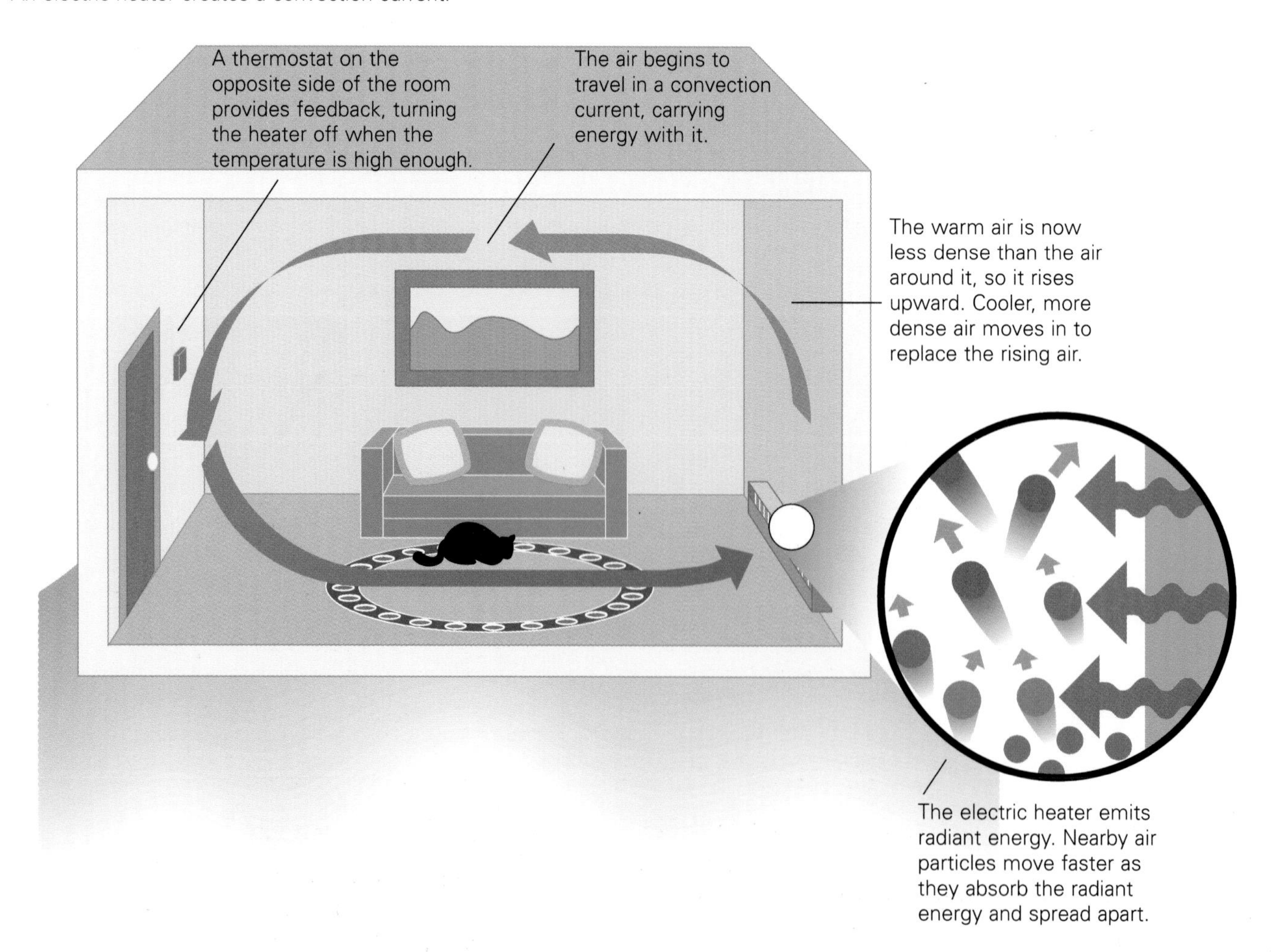

Hot-Water Heating Systems

Some homes and other buildings use hot water to provide heat. The water is under pressure, just like the water that goes to the taps in a home. **Figure 2** shows how hot water can be used to heat a home.

Figure 2

Some heating systems use hot water as a source of heat.

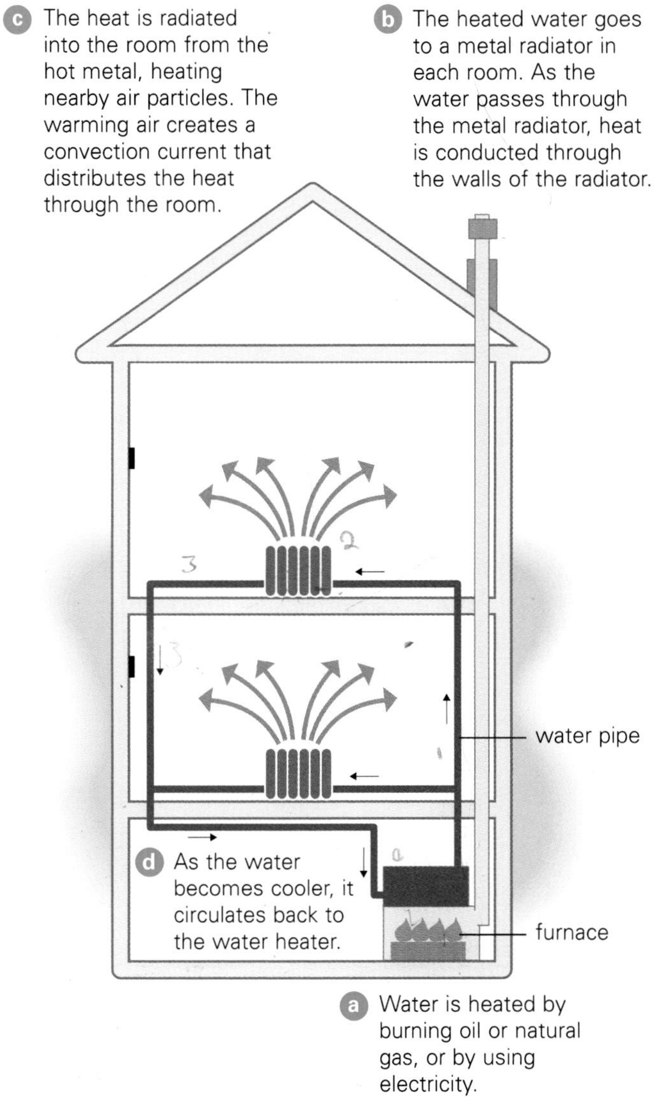

c The heat is radiated into the room from the hot metal, heating nearby air particles. The warming air creates a convection current that distributes the heat through the room.
b The heated water goes to a metal radiator in each room. As the water passes through the metal radiator, heat is conducted through the walls of the radiator.
water pipe
d As the water becomes cooler, it circulates back to the water heater.
furnace
a Water is heated by burning oil or natural gas, or by using electricity.

Forced-Air Heating Systems

In a forced-air heating system, air is heated in a furnace and then driven by a fan through ducts to the rooms. **Figure 3** shows a forced-air system.

Understanding Concepts

1. Where does the energy come from to heat the air in a single room in a hot-water heating system? in a forced-air heating system?
2. Using diagrams, compare the heating of air in a room to the development of a sea breeze. (6C)
3. How many methods of heat transfer are involved in a hot-water heating system? Give an example of each method.
4. Describe how a forced-air heating system works.

Making Connections

5. Two families, one with a hot-water heating system and the other with a forced-air system, leave for a one-week winter vacation. They turn down their thermostats to the same temperature to conserve energy. When they return, they increase their thermostats to the same temperature. Will the two homes heat up at the same rate? Explain your answer.
6. Describe the heating system used where you live. (6C) Use diagrams to show how the air is warmed, the method of heat transfer, and the feedback device that controls the system. Do you think the system uses energy efficiently? Explain.

Figure 3

A forced-air heating system

return duct
hot-air duct
fan
furnace

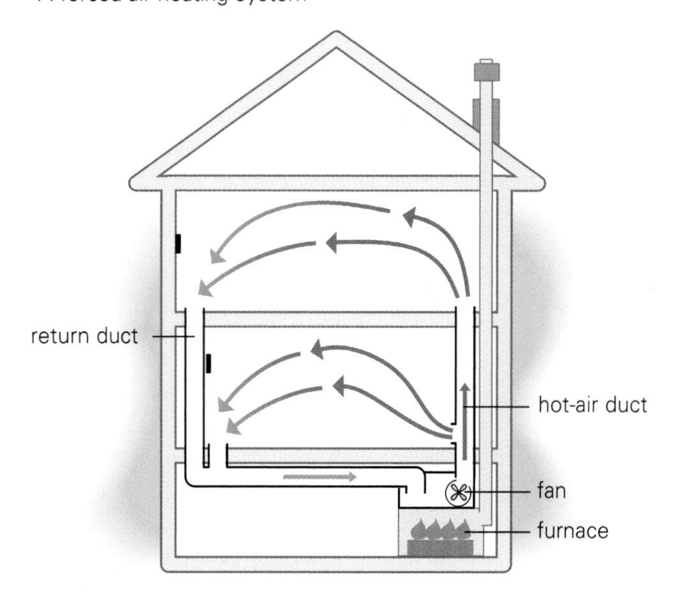

Design Challenge

Suggest a way to set up a feedback system to control the heat in the greenhouse challenge or the model pool challenge.

SKILLS HANDBOOK: (6C) Scientific & Technical Drawing

Wasting Heat

Heating bills are high for your school. This means that money that might be spent on books, computers, and sports equipment is being used to pay for heat. One way to reduce heating bills is to find out where warm air is escaping from the building.

To do this you could use an instrument called a thermal scanner. It measures invisible radiant energy. Then, with the help of a computer, the scanner changes the measurements into a "picture of heat" that you can see on a video screen. This process is called thermography. It helps show where heat is being lost from a building. **Figure 1** shows an example of a thermograph.

Figure 1

A thermograph. This house is losing lots of heat through its windows (white and orange areas), and a little through the front wall (green). The snow on the roof and the ground is cold (red).

Saving Heat

There are many ways to decrease loss of heat and other energy (**Figure 2**).

Figure 2

Some features that reduce heat loss from homes.

Source of heat loss / **Remedies**

walls and roofs

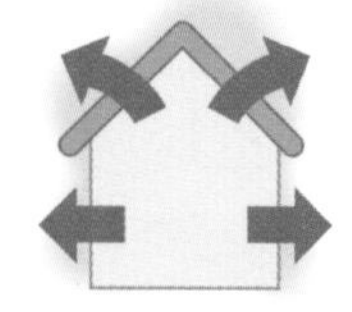

- Increase the amount of insulation in the basement walls, the roof, and the exterior walls.

air leakage

- Use a sealed air/vapour barrier to reduce air leakage and the buildup of moisture. The barrier is made of plastic and is placed on the inside of the insulated walls.

windows and doors
basement walls

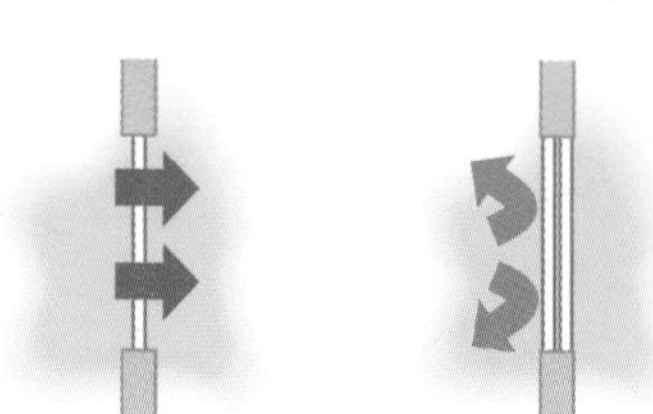

- Install windows that are either double- or triple-glazed, or use storm windows.
- Use doors that are made from good insulating materials or use a double-door system.
- Ensure that all windows and doors have tight weather seals.

lack of exposure to sunlight

- Whenever possible, face the home toward the Sun (east-west, with southern exposure) and use special designs to take advantage of solar energy.
- Use a screen of evergreen trees to protect the north side of the home from cold winds.
- Use a screen of deciduous trees to shade south-facing windows in the summer.

The Cost of Saving

Each energy-saving feature that is added to a building makes it more expensive. Many people cannot afford these features. And some people think that not all the features are necessary, or that they cost more than they save.

The Cost of Not Saving

We know that we must have heat for our buildings. But we also know that when we burn fuels, we add carbon dioxide to the atmosphere, increasing global warming, and we are using up a nonrenewable resource.

Understanding Concepts

1. **(a)** What are the advantages of adding heat-conservation features to a building?
   (b) How does each feature help save energy?
2. Why do we need to conserve energy?

Making Connections

3. Suggest another use for thermography and explain the benefits.

Design Challenge

Can you use any of the remedies for heat loss in your Challenge? Explain.

Debate Heat Conservation

Statement

All new buildings constructed in our community must include the latest energy-saving features.

Point

- We waste too much energy. We are also using up precious fossil fuels. Better insulation would reduce heat loss and the amount of fuel we use.
- Although adding features will cost more initially, it will save money in the long run through lower energy expenses.

Counterpoint

- We already have good building standards. By adding more standards, all we do is increase costs, making buildings too expensive.
- Energy-conserving features may not be cost effective. There are other, less expensive ways to conserve energy, such as turning off lights when they are not needed, or turning down the thermostat at night.

What do you think?

- Are you in favour of allowing only energy-conserving building standards in your community? Would you favour some energy-conserving features, but not others? Why or why not? Research the topic thoroughly and prepare to defend your opinions in a class discussion. (4A) (8D)

SKILLS HANDBOOK: (4A) Research Skills (8D) Exploring an Issue

2.16 Inquiry Investigation

SKILLS MENU
- Questioning
- Hypothesizing
- Planning
- Conducting
- Recording
- Analyzing
- Communicating

Controlling Heat Transfer

In our homes, schools, and other buildings, we like to keep the air temperature at a comfortable level. This requires energy—energy that costs money and that should not be wasted. To reduce the amount of energy needed to heat our buildings in the winter, we must learn how to control heat transfer. One way to do this is to make sure that our buildings have good insulation, as in **Figure 1**. Good insulators reduce heat transfer out of a building in winter and into a building in summer.

In this investigation, you will be testing the insulating abilities of various materials. Because convection and conduction transfer heat by the movement of particles, the best way to prevent heat transfer is to eliminate all particles—create a vacuum. A vacuum is difficult to create and maintain. The next best insulator is a substance that has particles that are well separated, like air. A region where air does not form a convection current is called a dead air space. Materials that include dead air spaces make good insulators. Think about where such spaces may exist when you write your hypothesis.

Question

Create a question for this investigation.

Hypothesis

1 Explain why the material you are testing should be a good or a poor insulator.

Experimental Design

2 You will design and carry out a controlled test of an insulator by measuring the rate of cooling of water in a container. The container should be surrounded by a single layer of the insulator you are testing.
- You will need a control container.
- Think about what variables you must keep constant and what variables you must change and measure.

3 Design your experiment, including a procedure, safety precautions, and a method for recording your data. (2E)

Materials

4 Create a list of materials you will use.

5 Have your teacher approve your design and materials list.

Procedure

6 Carry out your experiment and record your observations.

Figure 2
Which materials do you think are the best insulators?

SKILLS HANDBOOK: 2E Designing an Inquiry Investigation

Figure 1
Insulation prevents heat flowing, and so saves energy.

Making Connections

1. Which of the materials tested by the groups in your class would be best to wear to:
   (a) a Grey Cup game played in –5°C weather? Explain.
   (b) a baseball game in 30°C weather? Why?
2. A thermos bottle keeps cold liquids cold and hot liquids hot. Explain **Figure 3** and use a diagram to describe how a thermos bottle prevents conduction, convection, and radiation.

Analysis

7 To analyze your results, answer the following questions.

(a) Why was a control necessary in this investigation?

(b) Identify the dependent and the independent variables in your experiment.

(c) Create a graph of temperature vs. time for the insulator you tested. (7C)

(d) Compare your graph with the graphs of other groups who tested different materials.

(e) List the materials your class tested in order of best insulator to poorest.

(f) Use the particle theory to describe the feature(s) of the best insulators.

(g) How good was your prediction in your hypothesis? Explain any variation.

Figure 3
A thermos

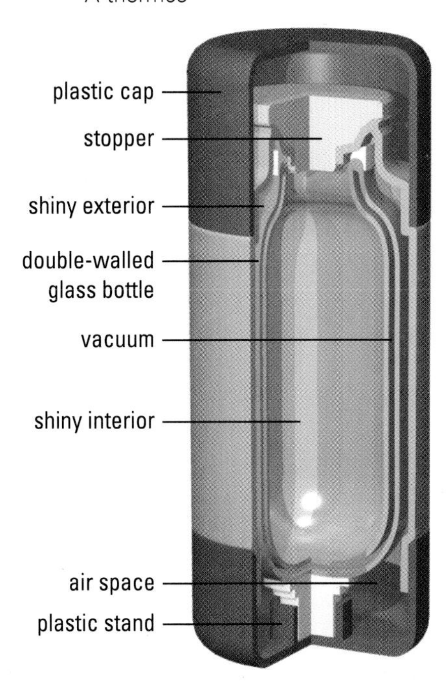

Design Challenge

Will your choice of material for the container in the swimming pool challenge affect how quickly the water can be heated?

Knowing some of the characteristics of a good insulator, should you change your choice of materials in the design of your pop can cooler?

(7C) Constructing Graphs

2.17 Inquiry Investigation

SKILLS MENU
- ○ Questioning
- ● Hypothesizing
- ○ Planning
- ● Conducting
- ● Recording
- ● Analyzing
- ● Communicating

Heating Various Liquids

From experience, you probably know that it is not smart to bite into an egg roll or a Jamaican patty that is fresh out of the oven, as shown in **Figure 1**. The outside may cool down fairly quickly, but the filling does not. Different substances cool down at different rates. They also heat up at different rates—but substances that heat up slowly also cool down slowly. Knowing that, which part of a Jamaican patty do you think should take longer to heat up—the pastry or the filling?

In this investigation, you will determine the heating rate of a liquid and then compare its heating rate with those of other liquids. What you learn can also be applied to solids and gases.

Materials

- apron
- safety goggles
- water
- vegetable oil
- glycerin
- hot plate (or other source of constant heat)
- Pyrex beakers
- support stand
- clamp
- thermometer
- stirring rod
- timing device
- balance
- tongs or insulating mitts

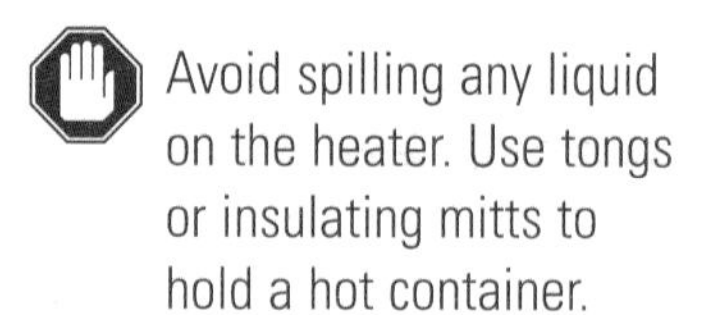

Avoid spilling any liquid on the heater. Use tongs or insulating mitts to hold a hot container.

Do not allow warm glycerin or vegatable oil to contact your skin. Do not breathe in over the warm liquids.

Question

Do all liquids change temperature at the same rate?

Hypothesis

1 (2C) Will 100 g of water need more time or less time than 100 g samples of vegetable oil or glycerin to go from 30°C to 60°C? Write a hypothesis for your prediction.

Experimental Design

This is a controlled experiment in which the mass and starting temperature of each liquid must be the same. You will measure the temperature of the liquid samples every 30 s as heat is added.

2 In your notebook, create a table to record data.

Procedure

3 (2A) Turn on the hot plate to medium so that it can heat up and reach a constant temperature while you are carrying out the next step.

4 Use a balance to measure 100 g of the liquid to be tested.

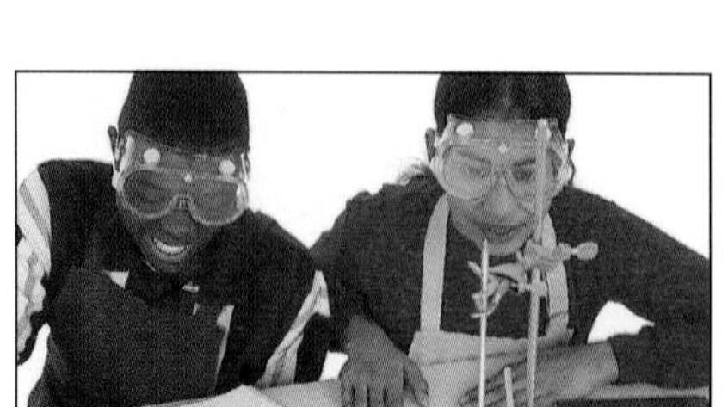

5 Place the beaker and liquid on the heater.
- As you stir with the stirring rod, observe the temperature of the liquid.
- When the temperature reaches 30°C, start taking readings of the temperature every 30 s until the temperature reaches 60°C.

(a) Record your observations.
- Carefully remove the beaker from the hot plate, using tongs or mitts. Your teacher will collect your liquid for proper disposal or storage.

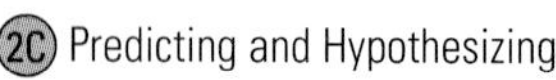

SKILLS HANDBOOK: 2C Predicting and Hypothesizing 2A Process of Scientific Inquiry

Figure 1

Why is the filling so much hotter than the pastry, since they both came out of the oven at the same time?

Making Connections

1. Which type of substance would likely make a better heat insulator, one that heats up rapidly, or one that heats up slowly? Explain.
2. **Figure 2** shows the results of an investigation comparing the heating rates of 1.0 kg samples.
   - **(a)** What can you conclude from this investigation?
   - **(b)** Which solid would be best for storing heat received from the Sun?
   - **(c)** Some cooking pots are made with copper bottoms and aluminum sides. Does this make sense? Explain.

Exploring

3. (7F) Describe how you would perform a controlled experiment to test your answer to question 1.

Analysis

6 Analyze your results by answering the following questions.

(a) (7C) Gather data for the other liquids from your classmates. In a line graph, plot the temperature–time points for your liquid, and then, in the same graph, for the other liquids. Use a different colour for each liquid.

(b) List the liquids in order from the longest to the shortest time it took each to heat from 30°C to 60°C.

(c) What variables were controlled in this experiment? Explain.

Figure 2

Data from an investigation

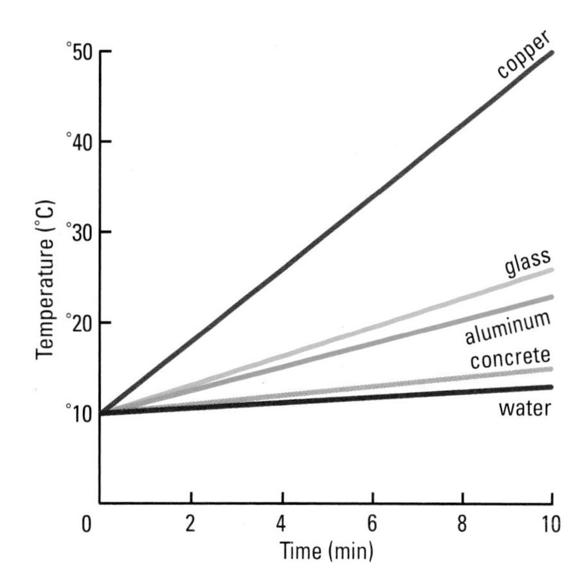

Design Challenge

When designing a control device that reacts quickly to temperature changes, should you choose materials that heat up rapidly or slowly? Could you use such a device in your Challenge?

Comparing Heat Capacities

Imagine you are at a beach by a lake on a clear summer morning. The water and sand temperatures are both at 20°C. As the Sun rises higher in the sky, you notice that the temperature of the water stays about the same, but the temperature of the sand rises fairly quickly. The water and the sand have different capacities to hold heat.

As you have learned, different substances heat up (and cool down) at different rates. The **heat capacity** of a substance is a measure of the amount of heat needed to raise the temperature of the substance; it is also a measure of how much heat the substance releases as it cools. The bar graph in **Figure 1** compares the heat capacities of several substances. Notice that water has a very high heat capacity.

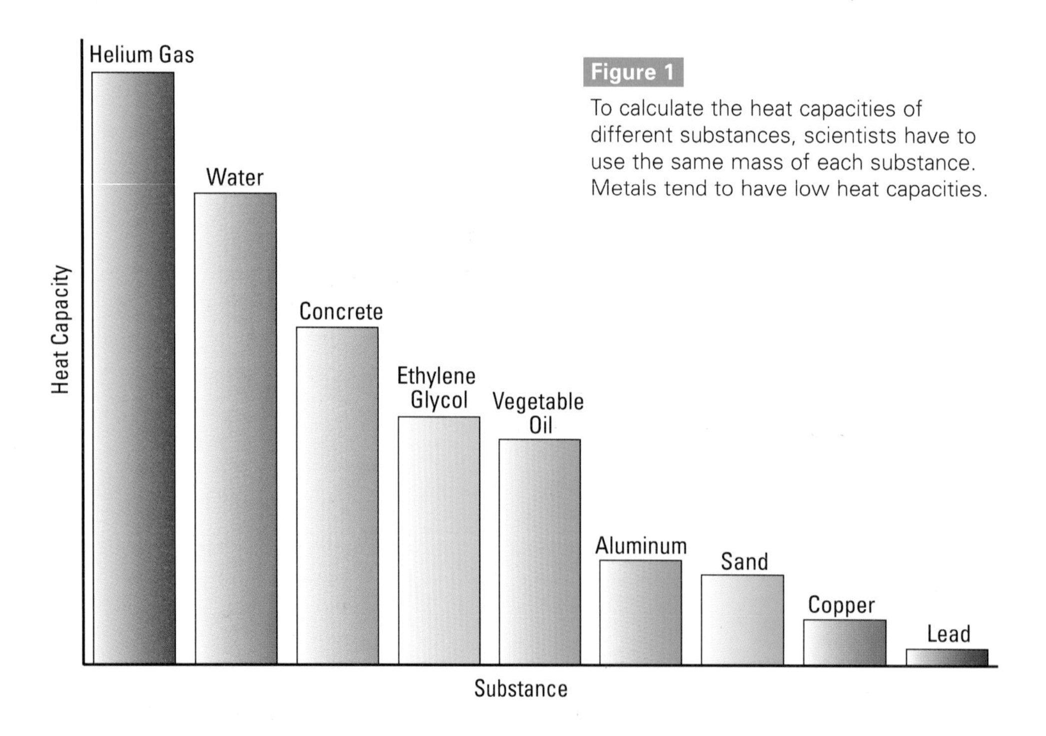

Figure 1
To calculate the heat capacities of different substances, scientists have to use the same mass of each substance. Metals tend to have low heat capacities.

How Does Water's High Heat Capacity Affect Climate?

Look at the map of Canada in **Figure 2**. The graph on the map shows the average January temperature in several cities. You might think Edmonton is so cold because it is farthest north. But why are Regina, Winnipeg, and Quebec City so cold? What makes Victoria, Toronto, and Halifax warmer in winter?

One thing the warmer cities have in common is that they are close to large bodies of water. As the weather gets colder in the fall and winter, the water takes a longer time to cool down than the land and the air do. Places such as Toronto are kept warmer by the water (Lake Ontario) than inland areas like London.

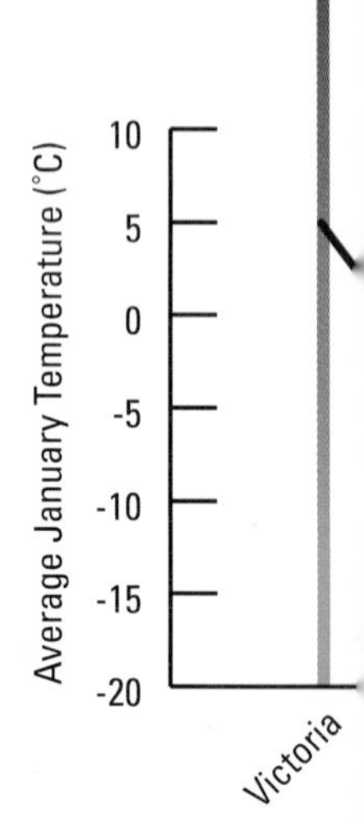

Heat Capacity and Solar Heating

How would your knowledge of radiant energy from the Sun and heat capacities of various substances help you design a solar-heated home? Look at the bar graph of heat capacities. Water and concrete have high heat capacities, so they would be good to store energy in a solar-heated home. Water could absorb radiant heat and become warm during sunny days. The water could be pumped to a storage tank, then energy in the water could be released during the night or when it is needed.

Concrete can be used in a similar way. As radiant heat from the Sun hits the concrete, the concrete slowly warms up. Then when night arrives, the energy stored in the concrete can keep the room warm as the concrete slowly releases heat.

Understanding Concepts

1. Assume you use a hot plate to heat equal masses of vegetable oil, copper, and aluminum from 20°C to 50°C. List the materials in order of fastest to slowest in reaching 50°C. Which has the highest heat capacity?
2. After a sunny day at the beach, which cools down more rapidly at night, the sand or the water? Explain.

Making Connections

3. You take a baked potato covered with aluminum foil out of a hot oven. You remove the foil, and very soon you can hold the foil in your hand, but not the potato. It remains hot for a long time. Explain why there is such a difference between the foil and the potato.

Exploring

4. (2E) Design an experiment to show that a substance that takes a long time to heat up also takes a long time to cool down.

Reflecting

5. What are some of the problems of harnessing solar energy in Canada?

Design Challenge

Based on what you have learned about the heat capacities of various substances, how could you modify your greenhouse and pool designs to prevent the internal temperature from dropping quickly at night? How could this information help you choose material for the design of your pop can cooler?

Figure 2

Average January temperature across Canada

SKILLS MENU
- ○ Questioning
- ○ Hypothesizing
- ○ Planning
- ● Conducting
- ● Recording
- ● Analyzing
- ● Communicating

Using Mechanical Ways to Produce Heat

Have you ever received a skid burn on your knees or elbows after falling on a gym floor? During the skid, heat was produced by the friction between the floor and your skin, as shown in **Figure 1**. Friction is a force that occurs between objects in motion that are touching. Friction can occur between solid objects. It can also occur between a fluid and a solid, such as a boat in water or the space shuttle in air as the shuttle lands.

Friction is just one example of a mechanical force, a force caused by objects in contact with each other. In this investigation you will study four mechanical ways of producing heat, including friction.

Materials

- apron
- safety goggles
- protective gloves
- metal coat hanger
- wooden block
- clamp
- hammer
- bicycle pump (or fire syringe)

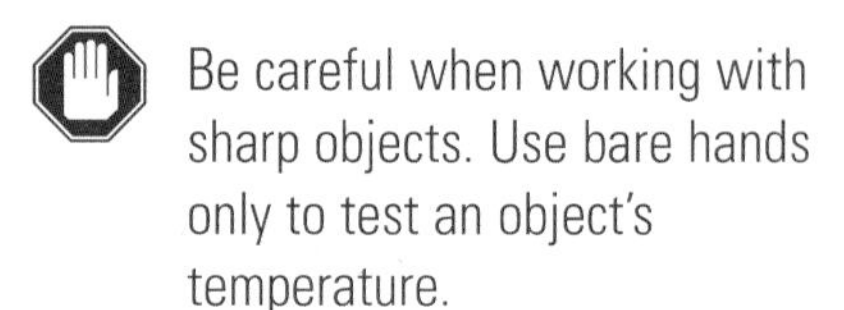

Be careful when working with sharp objects. Use bare hands only to test an object's temperature.

Make sure the wood is secure before pounding. Do not use your hand to secure the wood.

Question

What are some ways of producing heat using mechanical means?

Hypothesis

Mechanical forces can produce heat.

Experimental Design

In this investigation you will explore four ways of producing heat by mechanical means: friction, distortion (bending), percussion (pounding), and compression (squeezing).

Procedure

Part 1: Friction

1 Open your hands and hold your palms together. Rub your hands together about 20 to 25 times. Observe what happens when you press your hands closer together as you rub.

(a) Record your observations.

Part 2: Distortion

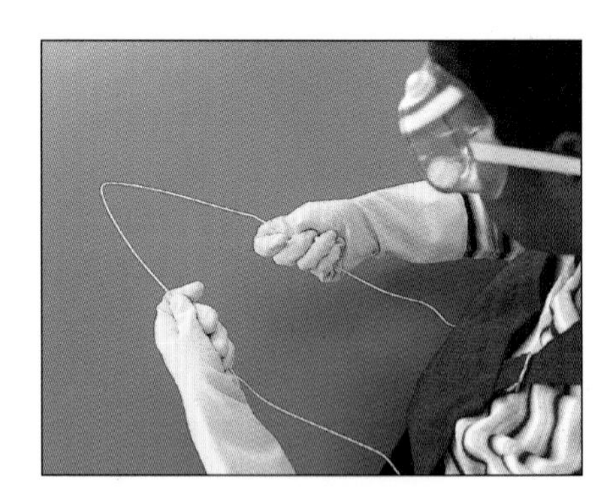

2 Use the palm of your hand to test the temperature of a metal coat hanger.

- Put on your gloves and bend the hanger back and forth quickly about 10 to 12 times.
- Test the temperature of the region that was bent.

(a) Record your observations.

SKILLS HANDBOOK: 2E Designing an Inquiry Investigation 2D Identifying Variables and Controls

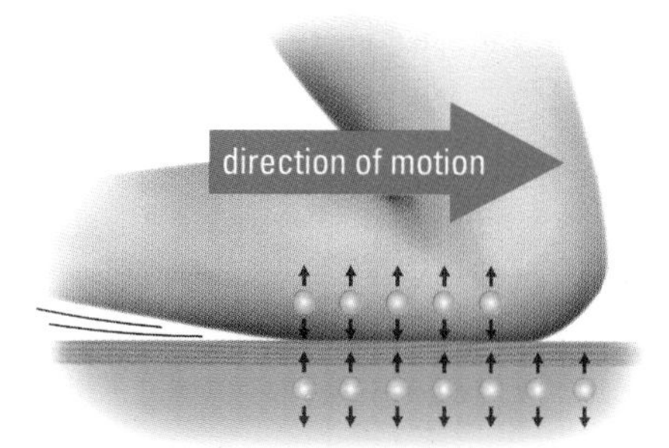

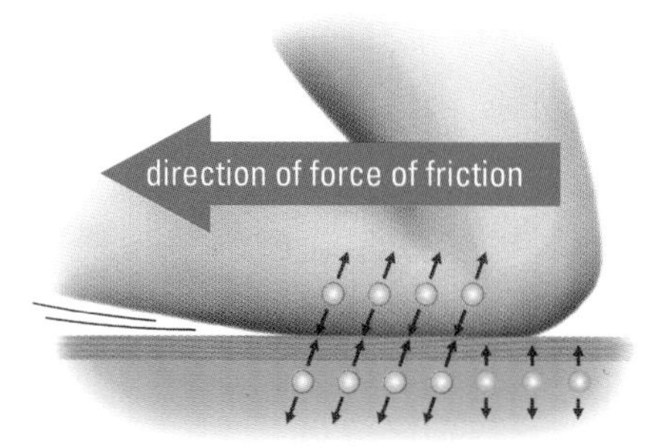

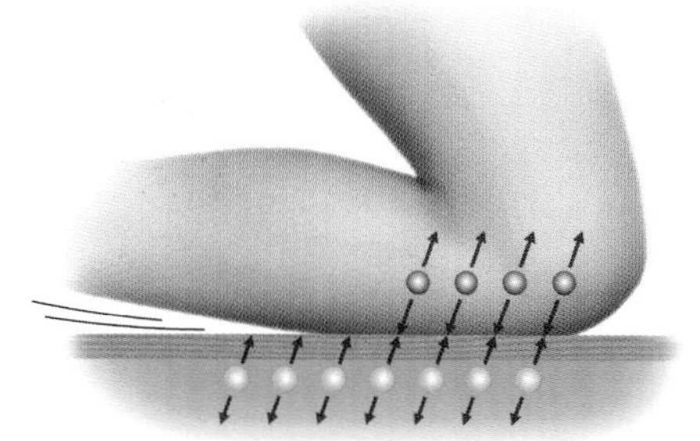

Figure 1

As two substances move past each other, the particles of one substance pull on the particles of the other substance. This causes resistance to motion and increases the motion of the particles in each substance, which increases the temperature.

Making Connections

1. Give two examples from outside your school of where you might find each of the four mechanical means of producing heat.
2. Motorcyclists often wear shiny, smooth leather, especially when riding at highway speeds. Explain why.
3. A dentist's drill requires water to be sprayed on to it when it is being used. Why is the water needed?

Exploring

4. Choose one of the mechanical ways of producing heat and design a controlled experiment to investigate it. What would be the independent variable? What variables would you have to control?

Part 3: Percussion

3 Use the palm of your hand to test the temperature of a wooden block.

- Put on your gloves and use the hammer to pound the same place on the block about 15 to 20 times.
- Test the temperature of the wood where you pounded it.

(a) Record your observations.

(b) Was heat the only type of energy you observed?

Part 4: Compression

4 Feel the valve of the bicycle pump and test its temperature.

- Compress the air in the pump by pumping hard.
- Test the temperature of the valve again.

(a) Record your observations.

Analysis

5 Analyze your results by answering the following questions.

(a) Use the particle theory to explain why the temperature of the surface of an object can increase because of friction.

Producing Heat

The Sun is our most important source of energy. However, energy from the Sun can be used to provide heat directly only in the daytime in clear weather. Even then the Sun's energy is not as convenient to use for heat as some other sources. An energy source is a material or method that provides energy we can use.

Energy sources can be grouped as renewable or nonrenewable. Renewable energy sources are those that are not destroyed in the process of being used; examples are solar energy and electricity produced from the moving water of rivers. Nonrenewable energy sources are gradually being used up and one day may run out entirely; examples are coal and oil. In the last few hundred years, people have used nonrenewable resources at an increasing rate.

We are now using wood more quickly than it can be replaced. Would you consider wood renewable or nonrenewable? As you read about examples of energy sources, think about whether they are renewable or nonrenewable.

Figure 1

The friction of spinning tires on the road surface produces lots of heat and some sound. The heat causes the air inside the tires to expand and inflate the tires. The energy transformation is: mechanical energy → heat and sound.

Heat from Mechanical Energy

In friction, distortion, percussion, and compression, mechanical energy changes into heat and perhaps some sound energy. In ancient times, people used friction between solid objects, such as two pieces of wood, to light fires. In modern times, we have created many more uses for friction (**Figure 1**).

Heat from Chemical Energy

Fuels such as oil, wood, coal, and natural gas have energy stored in them. This kind of energy is called chemical energy. Through the chemical process of burning, the energy is released as heat (**Figure 2**).

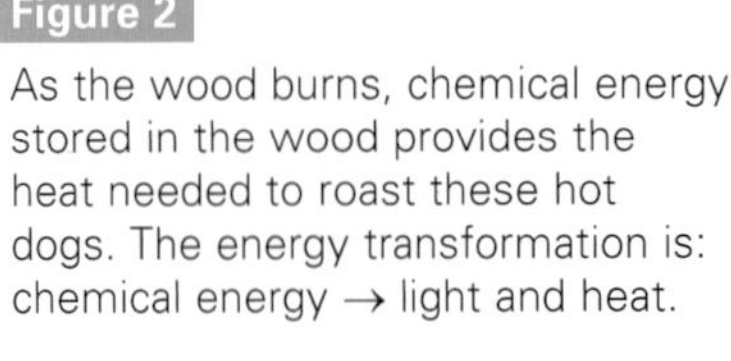

Figure 2

As the wood burns, chemical energy stored in the wood provides the heat needed to roast these hot dogs. The energy transformation is: chemical energy → light and heat.

Figure 3

Electricity flowing through this burner produces heat. The energy transformation is:
electrical energy → light and heat.

Heat from Electrical Energy

Electrical energy can change into heat, but first we must produce the electricity. There is more than one way to produce electricity. The energy in moving water can be used to produce electricity. It is also produced at generating stations by burning fuels such as coal and natural gas.

Once produced, electricity is sent through wires to our homes, where it is changed to heat in stoves, dryers, and many other appliances. For example, in a stove (**Figure 3**), an electric current passes through an element made of a material that resists the flow of electricity. The resistance causes the element to heat up. The greater the current, the hotter the element becomes until it is red hot (in a stove) or white hot (in a light bulb). Like other forms of energy, electrical energy ends up as heat.

Heat from Nuclear Energy

Tiny particles of materials are called atoms, and at the centre of each atom is the nucleus. Energy stored inside the nucleus is called nuclear energy. This energy can be changed into other forms of energy, such as heat, through processes called fusion and fission. In nuclear fusion, some small nuclei (plural of nucleus) join to form larger nuclei. In this process, some mass changes into energy. This is the way the Sun produces so much energy.

In nuclear fission, nuclei of a substance such as uranium split into smaller nuclei. Again, some mass changes into energy. Technologists use this process to produce heat at nuclear generating stations (**Figure 4**). The heat is used to create steam under pressure, and the steam causes large turbines to spin. As the turbines spin, both electrical energy and waste heat are produced. In both fusion and fission, the final form of energy produced is heat.

Figure 4

A nuclear generating station produces electrical energy from nuclear energy. The energy transformations are:
nuclear energy → heat (to heat water) → mechanical energy (of the spinning turbine) → electrical energy and heat.

Understanding Concepts

1. List three ways that heat can be produced. Indicate whether the method is renewable or nonrenewable. Give an example of how each is used in your daily life.
2. State two examples, other than those given here, in which heat is produced from:
   (a) mechanical energy
   (b) chemical energy
   (c) electrical energy
3. Where does the electrical energy come from for use in your home? in your school?

Making Connections

4. Tungsten is an important substance often used in light bulbs. Look at the table of melting and boiling points in section 2.7. Use the information there to explain why tungsten is a good substance to use in light bulbs.

Exploring

5. Write a science fiction story about a technological breakthrough in which a totally new and cheap source of heat is discovered. How would it affect our environment, business, and industry?

Reflecting

6. Someone once called heat the "graveyard" of forms of energy. Why is that a good description of heat?

Design Challenge

Describe how your Challenge design can be applied to help reduce our need for nonrenewable resources.

Heat Pollution

When you use an electric light bulb, only about 5% of the electrical energy becomes light energy. The rest becomes wasted heat. This is just one example of waste heat. All of our appliances, vehicles, factories, and buildings generate waste heat.

Because we use so much energy, there is more waste heat being produced than ever before. This waste heat is called **heat pollution**, because, like chemical pollution, it affects our environment.

Heat Pollution in and near Cities

The average air temperature in and near a big city is almost always a few degrees warmer than in the countryside several kilometres away. Buildings, roads, sidewalks, and cars absorb solar energy. Also, a lot of waste heat is produced by everything that uses energy, such as cars, trucks, heating, and lighting. Even the waste water from a city includes waste heat. The waste heat from all these factors combines to cause the air temperature in a city to rise.

In the summer, as the air gets hotter, air conditioning systems also produce waste heat. This creates a feedback loop: people react to the increase in temperature by turning up their air conditioning, which generates even more waste heat.

(a) Identify three sources of waste heat in the city in **Figure 1**.

(b) Explain how these sources produce waste heat.

Heat Pollution in Industries

Imagine how much water it would take to fill your science classroom. Now imagine a way to heat that much water so that in just five seconds its temperature goes up by 10°C! That is what happens at an electric generating station where fossil fuel (coal, oil, or natural gas) is burned to produce electricity.

At an electric generating station, like the one in **Figure 2**, about 60% of the chemical energy in the fossil fuel becomes waste heat.

Similar waste heat occurs at big industries, such as steel mills or pulp and paper mills, located along rivers and lakes. In a steel mill, for example, raw materials are heated in a blast furnace with temperatures as high as 1650°C. Large quantities of water are used to cool the finished metal.

(c) Why would an industry use water to cool heated materials rather than another substance?

(d) Which industries in your community would use a lot of water in their production process?

Figure 1

Waste heat from cities and industry raises the temperature of the air and of water in lakes and rivers.

Waste Heat and Wildlife

When hot water used in industry and electricity generation is discharged directly into a lake, it raises the temperature of the surrounding water, as shown in **Figure 1**. The temperature increase also reduces the water's oxygen content. This can disrupt the habitat of marine plants and animals. Living things are adapted to certain conditions in their habitat. Changes in habitat can be fatal. Some species, such as lake trout, need cold water and cannot survive increases in water temperature. If the water temperature rises, they must move out of the area or die. If other species that prefer warm water and require less oxygen can reach the warmer water, they may replace the cold-water species. Large-mouthed bass and pike are fish species that prefer warmer water.

(e) Suppose an electric generating station that has been running for years has to shut down for one week for repairs. What would happen to the water in a week?

(f) Describe what might happen to the surrounding marine life.

Waste Heat and Agriculture

Temperature increases can affect field crops and orchards. For example, the change could create a longer growing season. Changes in temperature might also affect precipitation patterns, for example by reducing the amount of snow that falls. Without snow cover the soil heats up more quickly when the sun shines. During a warm spell in winter, as the soil warms, plants and other organisms in the soil might be fooled into thinking it is spring. Snow is also a source of ground water in the spring, which plants can draw on with their roots.

(g) How would a longer growing season and changes in precipitation patterns affect a farmer's choices of what to grow?

(h) What might happen if plants and organisms are fooled into thinking it's spring?

Figure 2

ⓐ Steel Mill
Steel mills pump their hot coolant water into ponds, where the waste heat escapes into the air. The pond water is cleaned and reused.

ⓑ Electric Generating Station
Every second, more than 30 000 L of water from the lake pass through this electric generating station. The water takes away the waste heat and comes back to the lake almost 10°C hotter.

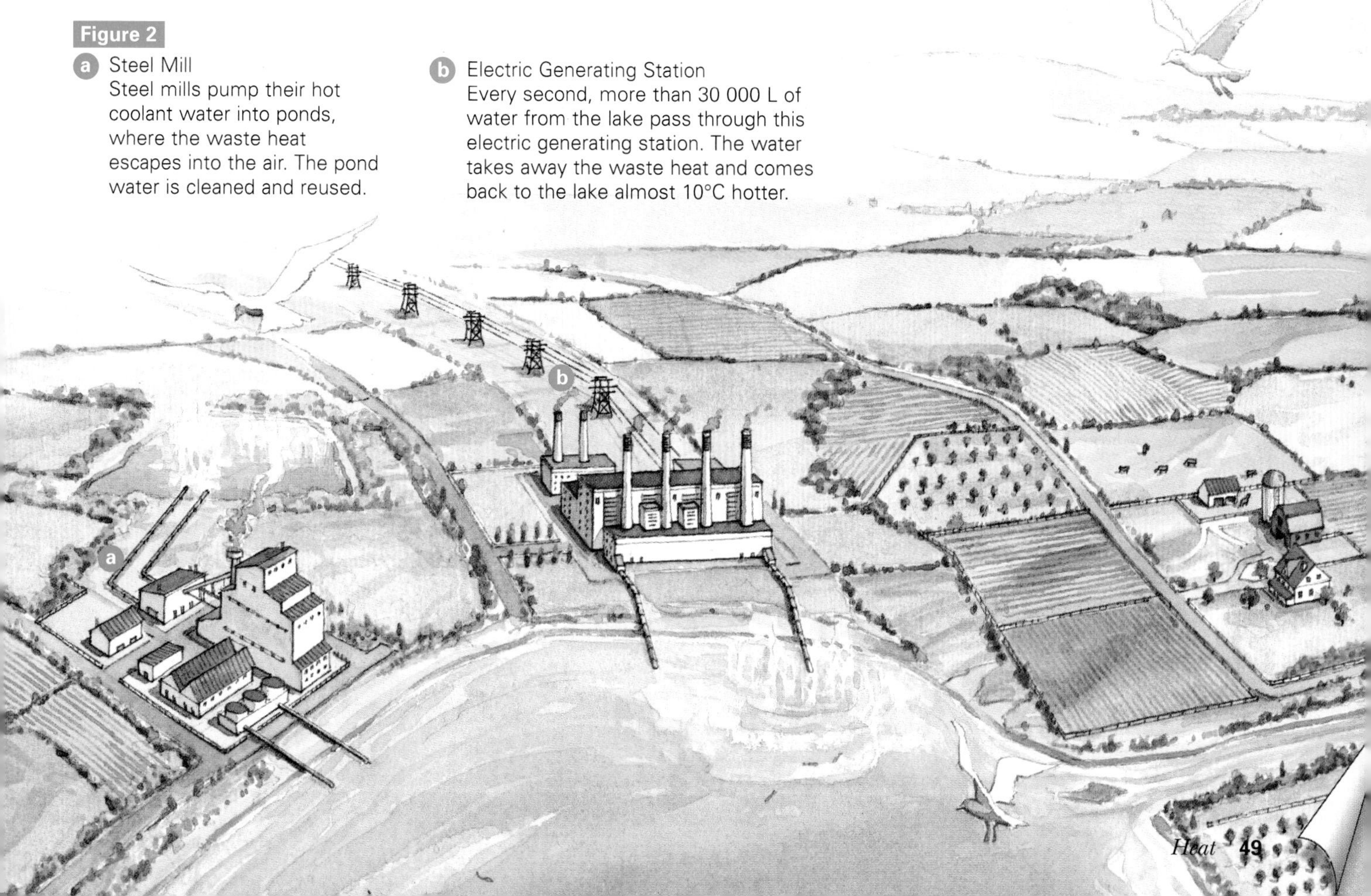

How Much Energy Is Wasted?

An incandescent light bulb has an efficiency of only 5%, as shown in **Table 1**. This means only 5% of the energy input (electricity) becomes useful output (light). The other 95% becomes waste heat. Knowing the efficiencies of various devices may help you decide which device to use.

(i) Look at the efficiencies of the devices in **Table 1**. In each case, what percentage of the energy used is wasted heat?

(j) Why is an electric heater rated at 100% efficiency?

Table 1: Efficiencies of Some Devices

Device	Efficiency
incandescent light	5%
fluorescent light	20%
automobile	25%
fossil-fuel electric generating station	40%
wind generator	55%
gas furnace	85%
falling-water generating station	95%
large electric motor	95%
electric generator	98%
electric heater	100%

Reducing Waste in Mechanical Systems

The moving parts of any mechanical system add to heat pollution, mainly because of friction. This happens in machinery, in motors of all sizes, and in all vehicles. The friction of moving parts, such as the wheels and motor parts, is reduced by making the parts smooth and using a substance that reduces friction, called a **lubricant**. Oil is a lubricant.

Another form of friction, called air resistance, pushes against moving vehicles. Cars, trucks, and other vehicles can be designed to reduce air resistance, for example, by using a curved shape that goes through the air smoothly.

Reducing friction and air resistance helps to reduce heat pollution, as the less friction in the moving parts of a vehicle, the less fuel the vehicle must burn to operate.

(k) Look at the vehicles in **Figure 3**. What features of each model would make it more efficient in using the chemical energy from gasoline?

(l) What other methods could be used to reduce the amount of gasoline used by vehicles?

Figure 3
Two possible vehicles

Figure 4
Commercial greenhouses could use hot water from a generating plant to help maintain a constant temperature.

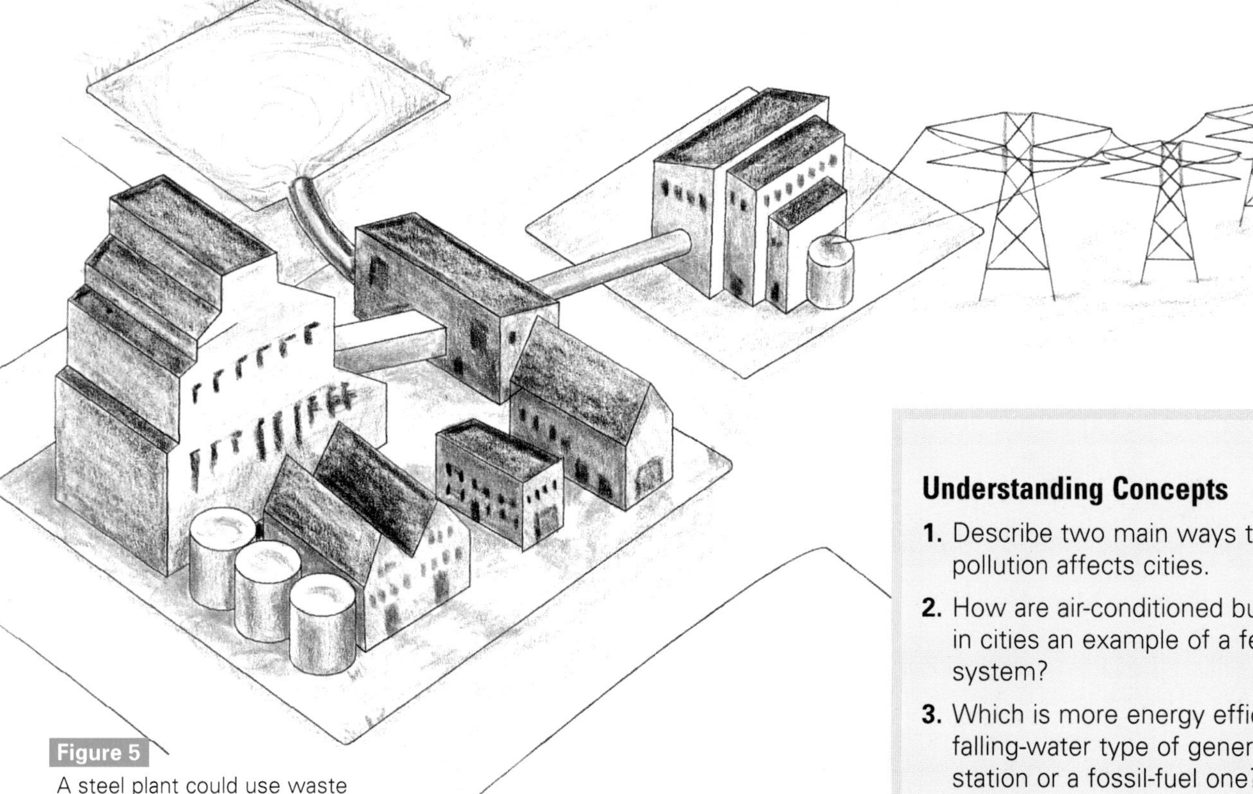

Figure 5
A steel plant could use waste heat to generate electricity.

Recycling Waste Heat

Waste heat doesn't need to be wasted. Heat from generating stations could be used in industrial processes and to heat nearby buildings. The process of providing electricity and heat at the same time is called **cogeneration**. **Figure 4** shows how waste heat from an electric generating station can be used by a nearby industry. **Figure 5** shows how an industry can burn fuel to provide the heat it normally needs and use any extra heat to produce electricity that can be sold to the local power company.

(m) Suggest another way waste heat from an electric generating station could be used.

(n) It is possible to heat homes using waste heat from generating stations or industry, but the homes would have to be close to the source of heat. List some economic, social, and environmental advantages and disadvantages of using waste heat in this way.

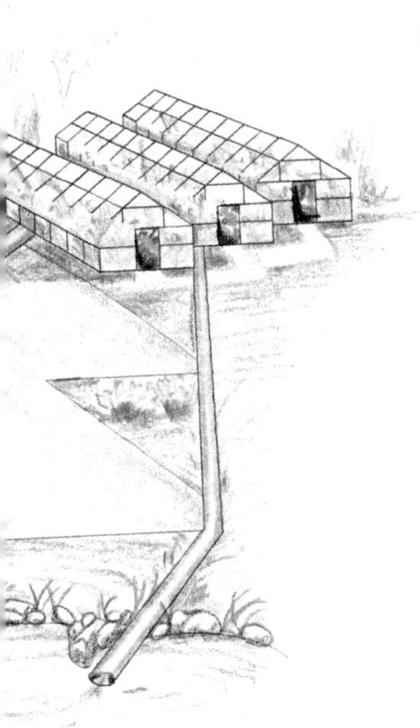

Understanding Concepts

1. Describe two main ways that heat pollution affects cities.
2. How are air-conditioned buildings in cities an example of a feedback system?
3. Which is more energy efficient, a falling-water type of generating station or a fossil-fuel one? Explain.

Making Connections

4. If you were designing a car, what could you do to ensure that friction and air resistance were kept to a minimum? (3D)
5. **(a)** Waterfowl (birds that spend much of their time in or on the water) need open, unfrozen water to survive. How could waste heat from cities affect the habitat and behaviour of waterfowl?
   (b) Most waterfowl eat fish or small animals that feed on water plants. How might the change in waterfowl behaviour affect other living things near the city? Explain.

Design Challenge

Will your greenhouse or swimming pool model generate excess heat? What could you add to your design to recycle some of that extra energy for other purposes? What could you do with your pop can cooler to keep it from being exposed to waste heat?

SKILLS HANDBOOK: (3D) Planning a Prototype

Solar Heating

Solar energy is important for life on Earth. It may become even more important to us as our need for energy increases. Solar energy can be used to heat places where we need to warm water or air, such as homes, larger buildings, and swimming pools.

There are two basic ways of using solar energy to provide heat. One way is passive; the other is active. As you read about these ways, think about why the passive way is much cheaper to set up than the active way.

Passive Solar Heating

In passive solar heating, the word "passive" means that the system lets the solar energy in and prevents much heat from getting out. Passive solar heating is not expensive and is easy to maintain. Most homes use passive solar heating simply by allowing sunlight to shine in the windows. **Figure 1** shows a house where passive solar heating has been planned.

The Greenhouse Effect

Windows are important in solar-heated homes as well as in other structures, such as greenhouses. The **greenhouse effect** is the process of trapping radiant heat inside a structure. The name came from the greenhouse, where glass is used to trap heat. The greenhouse effect is shown in **Figure 2**.

Figure 1
Window treatment and building materials are very important in a home designed to make best use of passive solar heating.

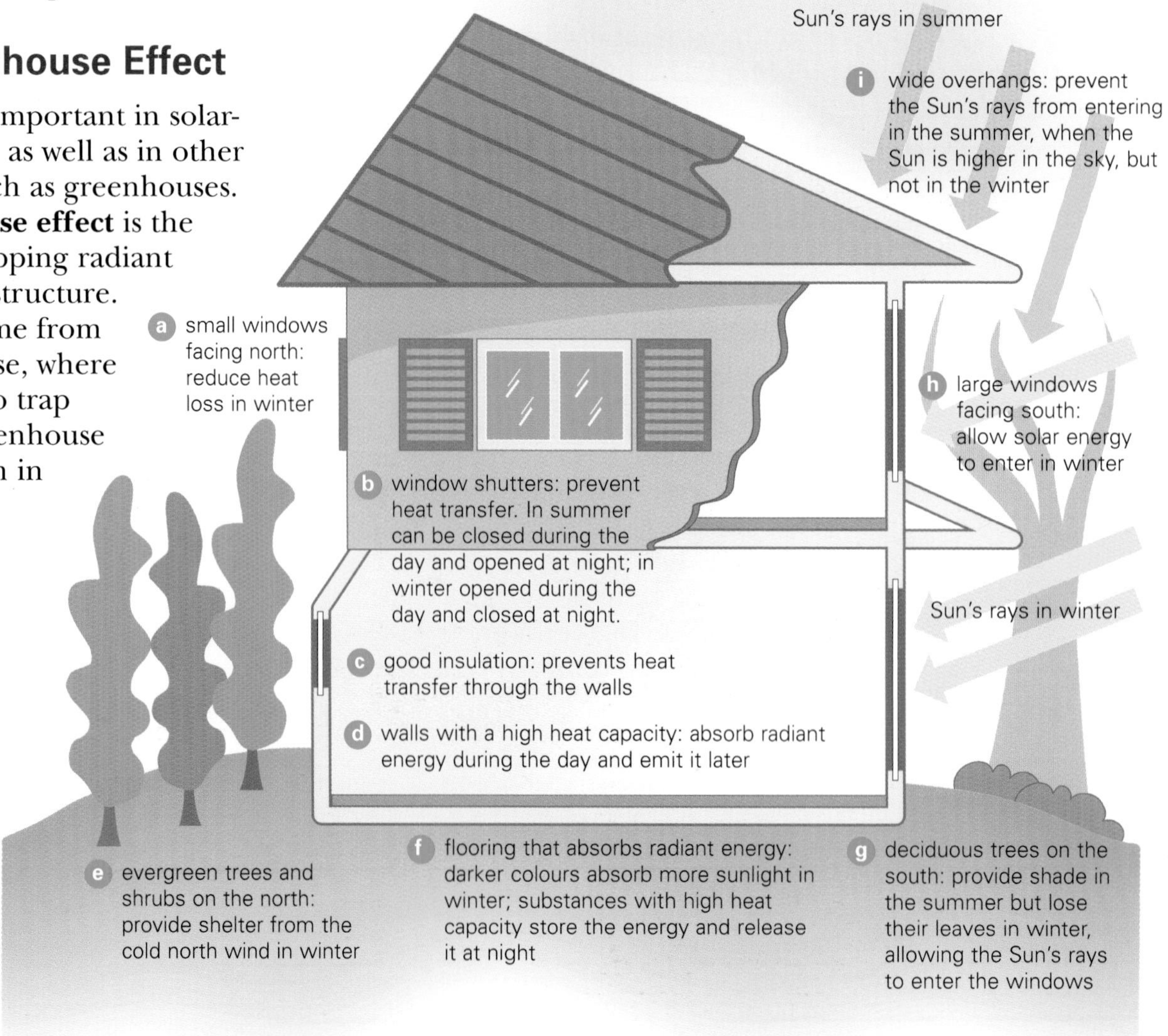

Sun's rays in summer

(i) wide overhangs: prevent the Sun's rays from entering in the summer, when the Sun is higher in the sky, but not in the winter

(a) small windows facing north: reduce heat loss in winter

(h) large windows facing south: allow solar energy to enter in winter

(b) window shutters: prevent heat transfer. In summer can be closed during the day and opened at night; in winter opened during the day and closed at night.

Sun's rays in winter

(c) good insulation: prevents heat transfer through the walls

(d) walls with a high heat capacity: absorb radiant energy during the day and emit it later

(e) evergreen trees and shrubs on the north: provide shelter from the cold north wind in winter

(f) flooring that absorbs radiant energy: darker colours absorb more sunlight in winter; substances with high heat capacity store the energy and release it at night

(g) deciduous trees on the south: provide shade in the summer but lose their leaves in winter, allowing the Sun's rays to enter the windows

Figure 2

The greenhouse effect

(a) Radiant energy from the Sun is made up mainly of waves that can travel through glass.

(b) Plants and other objects in the greenhouse absorb the radiant energy. This causes their temperature to increase.

(c) At the higher temperature, the objects emit their own radiant energy, but it is composed of waves that cannot get through the glass. This trapped radiant energy helps keep the greenhouse warm, even in winter.

Active Solar Heating

Another way to heat buildings is active solar heating. The word "active" means that the system absorbs as much solar energy as possible and distributes it throughout the building. **Figure 3** shows an active system. An active solar-heating system usually requires another source of energy besides the Sun, at least as a backup during times when there is little sunshine.

Figure 3

One type of active solar-heating system. In the solar collector for this system, radiant energy passes through a clear plastic or glass covering and is absorbed by a dark-coloured collector plate.

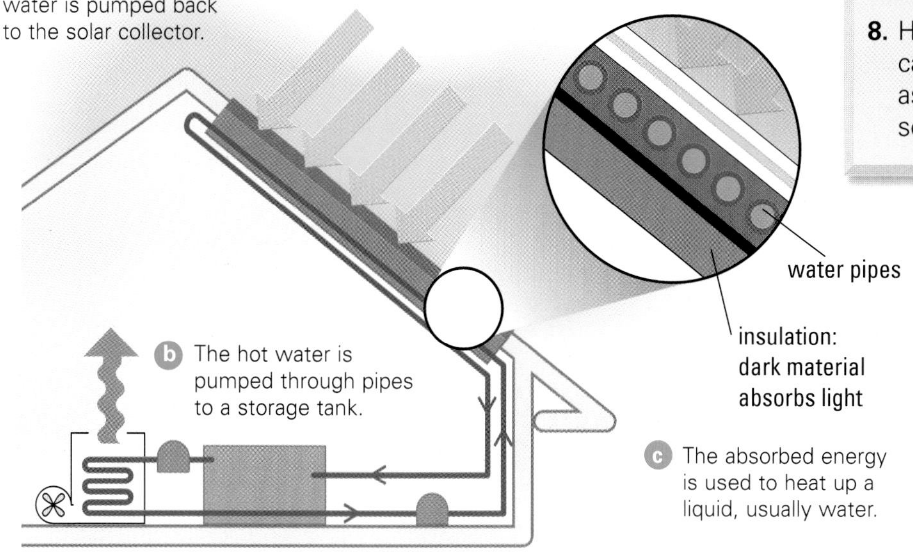

(a) Heat from the water is transferred to the air, which is pumped through the house using a forced-air system. The water is pumped back to the solar collector.

(b) The hot water is pumped through pipes to a storage tank.

(c) The absorbed energy is used to heat up a liquid, usually water.

(d) Radiant energy from the Sun is absorbed by a solar collector on the roof. A steep roof allows the Sun's rays to hit more directly in winter months.

water pipes

insulation: dark material absorbs light

Understanding Concepts

1. List several advantages of passive solar heating in homes.
2. How is the greenhouse effect used in passive solar heating?
3. Figure 3 shows the water storage tank in the basement. Would the attic be a better location for this tank? Why or why not?
4. What is the function of the insulation in the solar collector?

Making Connections

5. What special considerations would be important in a solar-heated home in your own region? Assume no other sources of heat are allowed.
6. Draw diagrams showing how solar heating could be used in two of the following places: apartment buildings, schools, hospitals, barns, industrial buildings. (6C)

Exploring

7. Describe and evaluate solar heating in your area.
   - **(a)** Research how many hours of sunshine there are on an average day in midsummer and midwinter. (4A)
   - **(b)** Describe how climate affects the type of solar heating suitable for buildings in your area.

Reflecting

8. How does knowing about the heat capacity of various substances assist an architect in designing a solar house?

Design Challenge

Would an active or a passive solar system work best in the Challenge design for the swimming pool? for the greenhouse?

SKILLS HANDBOOK: (6C) Scientific & Technical Drawing (4A) Research Skills

Design Challenge

SKILLS MENU
- Identify a Problem
- Planning
- Building
- Testing
- Recording
- Evaluating
- Communicating

Design and Build a Device That Controls or Uses Heat

All life on Earth depends on heat. If plants and animals get too hot or too cold, they cannot survive. In nature, we can find many examples of animals that keep a healthy temperature by controlling heat transfer. Humans use what they learn from nature to develop technologies to control heat.

Figure 1 Warm pop isn't as pleasant as cold pop—but sometimes a fridge isn't handy.

1 A Device That Delays Heat Transfer

Problem situation

We use appliances to cook food and to keep it cool. Ovens, refrigerators, and similar devices will not waste nearly as much energy if they are designed to prevent heat transfer to or from the outside.

Design brief

- Design a container that will keep a cold pop can as cool as possible from breakfast until lunchtime.

Design criteria

- The container must fit inside a standard lunch box.
- The pop can should be 355 mL. It should start at refrigerator temperature.

2 A Swimming Pool Heated by the Sun

Problem situation

As the world's population increases and our standard of living rises, we use more energy. But many of our sources of energy are not renewable, for example oil and gas. Using oil or gas heaters to warm a swimming pool is wasteful.

Design brief

- Design a controllable device that uses only energy from the Sun to heat the water for a model swimming pool to 25°C.

Design criteria

- The device should work either in sunlight or light from a bright lamp.
- The model pool must hold at least 250 mL of water, all of it usable for swimming.
- The water must be cold water from the tap.
- The device must be safe for swimmers.

Figure 2 A lot of energy is needed to heat water in a swimming pool.

3 An Environment to Protect Plants

Problem situation

Every year, more people in Canada discover the advantages of growing their own plants for food or simply for beauty. But the growing season in Canada is short, because many plants die when struck by frost.

Design brief

- Design a greenhouse or other environment that would allow a plant to grow even when outdoor temperatures are low. The environment must also resist overheating, to prevent plants from dying from too much heat.

Design criteria

- The environment must support at least one plant.
- To avoid energy waste, only energy from the Sun, either direct or stored, can be used.
- The temperature in the environment should remain fairly constant. It should not drop below 10°C, even if outside temperatures are lower than 10°C for five hours. It should not rise above 30°C, even in bright sunshine on a warm day (25°C).

Figure 3
This greenhouse protects tropical plants from the outside cold.

Assessment

Your model will be assessed according to how well you:

Process
- Understand the problem
- Develop a safe plan
- Choose and safely use appropriate materials, tools, and equipment
- Test and record results
- Evaluate the sustainability of your model, including cost, health, and social implications, as well as suggestions for improvement

Communicate
- Prepare a presentation
- Use correct terms
- Write clear descriptions of the steps you took in building and testing your model
- Explain clearly how your model solves the problem
- Make an accurate technical drawing for your model

Produce
- Meet the design criteria with your model
- Use your chosen materials effectively
- Construct your model
- Solve the identified problem

When preparing to build or test a design, have your plan approved by your teacher before you begin.

Unit 2 Summary

Reflecting

- Reflect on the ideas and questions presented in the Unit Overview and in the Getting Started. How can you connect what you have done and learned in this unit with those ideas and questions? (To review, check the sections indicated in this Summary.)
- Revise your answers to the Reflecting questions in ❶, ❷, ❸ and the questions you created in the Getting Started. How has your thinking changed?
- What new questions do you have? How will you answer them?

Understanding Concepts

- describe how thermometers operate using expansion and contraction 2.3
- explain matter and moving particles 2.5, 2.7, 2.8, 2.10
- distinguish between temperature and heat 2.5
- observe energy transfers from hotter objects to cooler ones by convection, conduction, or radiation 2.5, 2.8, 2.9, 2.10, 2.12
- describe how adding or removing heat from a substance may change its state 2.6, 2.7
- observe how substances heat up and cool down at different rates, depending on their capacity to store heat 2.17, 2.18

Applying Skills

- create a scale of temperatures, using experience and deduction 2.1
- plan and conduct an investigation to identify the property of matter that explains how a thermometer works 2.2
- describe and investigate the characteristics and changes of the states of matter 2.5, 2.6, 2.7
- observe and experiment with the properties of a convection current 2.8
- design, plan, and carry out an investigation to determine which solids are the best heat conductors 2.10
- communicate ideas, procedures, and results of investigations of heat transfer by convection, conduction, and radiation 2.8, 2.10, 2.12
- investigate and explain the use of insulating materials to control heat transfer 2.15, 2.16

- investigate and compare the heat capacities of various substances 2.17, 2.18
- classify mechanical, chemical, electrical, and nuclear ways of producing heat and indicate whether they are renewable or nonrenewable 2.19, 2.20

Making Connections

- observe how plants and animals are sensitive to changes in temperature and depend on heat for survival 2.4
- describe how convection currents affect weather 2.9
- identify some careers that require knowledge of heat and temperature 2.11
- explain how Earth's water cycle is a series of heat transfers 2.13

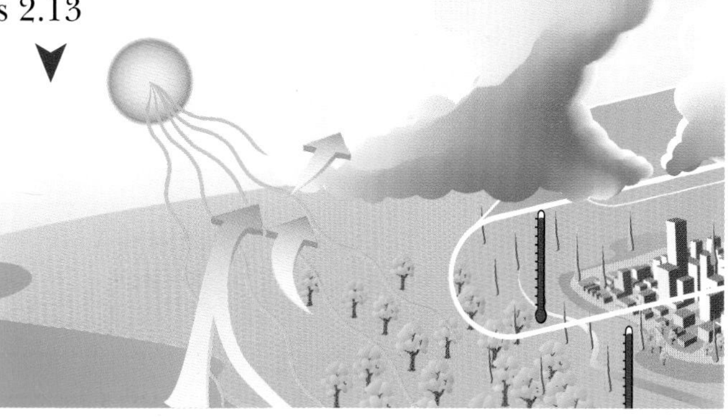

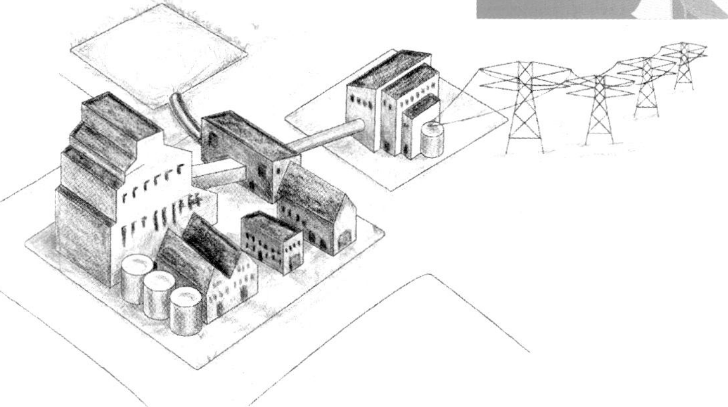

- explain how homes can be heated with hot-water systems or forced-air systems 2.14
- identify ways that waste heat can be reduced through energy-saving measures and recycled to reduce the amount of heat pollution 2.15, 2.21
- describe the use of solar energy to heat buildings, using either passive or active systems 2.22

- understand and use the following terms:

boiling point	heating curve
cogeneration	liquid
conduction	lubricant
contraction	melting point
convection	particle theory
convection current	radiant energy
expansion	radiation
gas	solid
greenhouse effect	temperature
heat	thermocouple
heat capacity	thermometer
heat conductor	thermostat
heat pollution	

Unit 2 Review

Understanding Concepts

1. When Marco first enters a swimming pool, he thinks the water is cold. After a couple of minutes in the pool, he thinks the temperature is just fine. Why does he change his mind?
2. Describe how a liquid thermometer and a thermocouple are the same and how they are different.
3. Vegetables are being cooked in a pot on the stove. Use the particle theory to explain why the lid starts to jump up and down.
4. A boy who feels ill touches his forehead with his hand to see if he has a fever. Will he be able to tell? Explain.
5. Explain why a liquid clinical thermometer has
   (a) a small range of Celsius degrees;
   (b) a narrow bore just above the bulb.
6. Describe situations where a solid thermometer would be more useful than a liquid one.
7. Describe how you and your class could act out Brownian motion.
8. Use the particle theory to explain expansion and contraction of a solid when its temperature is changed.
9. Is it possible to add heat to a material without changing its temperature? Explain.
10. Identify the three states of matter and give two examples of substances in each state.
11. Use the particle theory to explain why water expands when it changes from the liquid state to the gas state.
12. In which of the three states of matter does convection occur? Why can convection not occur in the other state(s) of matter?
13. A magician at the fall fair had five copper pennies like those in **Figure 1**, each from a different year. She put the pennies into a hat and asked someone in the audience to pick out one coin. Then she said, "Pass the coin around so everyone can see the year on it. Then put the coin back into the hat, and I'll try to pick out the same coin." When she reached into the hat, there was a brief pause and then she pulled out the correct penny. How did she do it?

Figure 1

14. What happens to the average energy of the particles in milk when the milk is taken from the refrigerator and begins to warm up? Use the concept of heat transfer and the particle theory to explain your answer.
15. What colour of clothing should be worn to keep cool on a hot, sunny day in summer? Explain your answer.
16. Give some examples of materials that slow heat transfer. In each case, state an example of where the material is used.
17. Describe how mechanical energy can be changed into heat.
18. State the type of energy that is the source of heat in each situation described below.
    (a) A dentist's high-speed drill becomes hot when drilling teeth.
    (b) Ancient people used flint, a very brittle type of stone, to start fires.
    (c) Water from Earth's oceans and lakes evaporates, forming clouds.

19. Explain why wearing loose clothing on a cold day might provide good insulation. What else might be needed?

20. A down jacket temporarily loses some of its insulating ability when it becomes wet.

(a) Explain why this occurs.

(b) What would you do, and why, to restore much of the jacket's insulating ability?

21. Which is better for keeping food and drinks cool in a picnic cooler: water at 0°C or ice at 0°C? Explain your answer.

22. Look at the experimental setup in **Figure 2**. The water in both beakers starts at the same temperature and the hot plates produce the same amount of energy. Will the water boil first in the beaker on the left or on the right? Use the words "particles," "average energy," and "heat" in your explanation.

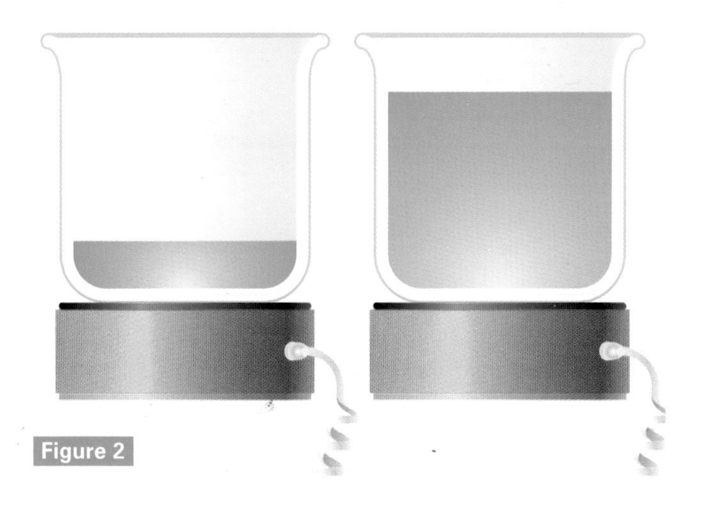

Figure 2

Applying Skills

23. Explain the significance of these temperatures:

(a) 100°C (b) 37°C (c) 0°C

24. Make up a poem or short story about heat and temperature using words in the vocabulary list in the Unit Summary.

25. Describe how a bimetallic strip could be used to make a thermometer. How would you calibrate it? Would it be very accurate? Explain.

26. Suppose you are helping someone find a house to buy.

(a) What questions related to heat could you ask to be sure the buyer is getting an energy-efficient home?

(b) Identify the steps that you could take to find answers to the questions in (a).

27. Describe a controlled experiment to determine which method of cooking corn or potatoes would require the least amount of energy.

28. A consumer magazine hires you to test kettles used to heat water to see which should be recommended.

(a) What would you test?

(b) How would you perform a controlled investigation to test those factors?

(c) Design data tables for the tests.

29. In an investigation equal masses of two liquids were allowed to cool from the same starting temperature. The results are shown in **Figure 3**. From the graphs, describe what you can about:

(a) the change of state of each liquid

(b) the heat capacity of each liquid

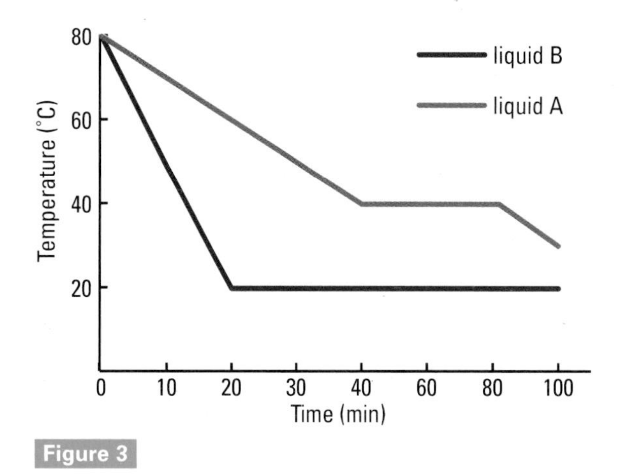

Figure 3

30. Design a home for a tropical climate. Include reasons for the design features.

31. Plan an investigation to determine how good snow is as an insulator.

32. Using terms you've learned in this unit, design a concept map that illustrates heat transfers.

Making Connections

33. State one useful application for each of the observations listed below:

(a) Liquid mercury expands when heated and contracts when cooled.

(b) Metals conduct heat better than glass does.

(c) Some metals expand more than others when heated.

34. How does thermal underwear help prevent the loss of body heat?

35. Use the particle theory to explain why steam causes a more serious burn than hot water.

36. If you are planning a winter hiking trip, how will you keep warm and dry on the hike?

37. In winter, many people turn down the heat in their homes before going to bed. Then they may hear the floors start to creak. Why does this creaking occur?

38. Although snow is cold, it can act as a very good heat insulator.

(a) Explain why snow is a good insulator.

(b) Describe how the insulating properties of snow are useful for plants, animals, and people in Canada's north.

39. Look at the paved surface of a highway bridge in **Figure 4**. Which diagram represents the paved surface in the summer? Explain your answer.

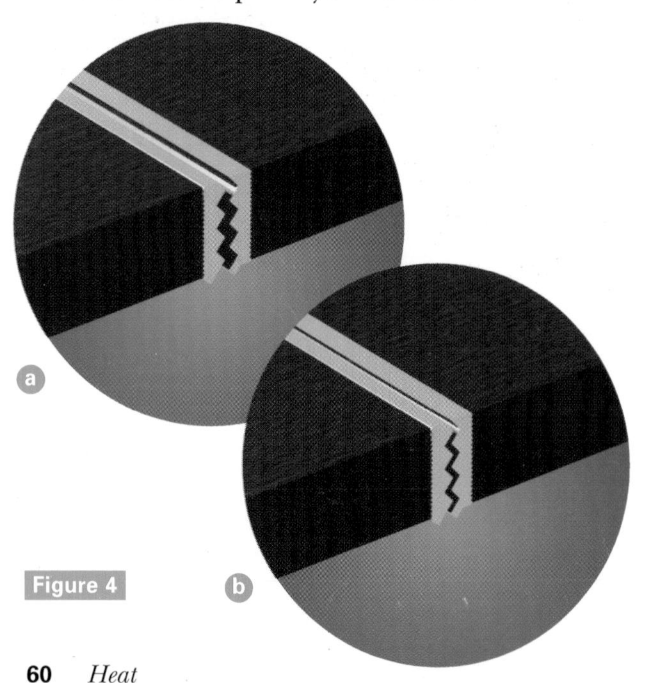

Figure 4

40. How would you restore a dented table tennis ball it to its original shape?

41. **Figure 5** shows that steel rods are used in concrete to strengthen the concrete. Fortunately, steel and concrete expand the same amount when heated. Describe what might happen if they didn't.

Figure 5

42. Explain why each of the following are concerned about heat transfer:

(a) an architect

(b) a hot-air balloonist

(c) a chef in a restaurant

(d) a manufacturer of winter boots

(e) a long-distance runner

43. Newspapers sometimes refer to solar energy as "free." Is it really "free"? Explain what expenses you would have in converting your home to active solar heating.

44. Using what you know about the properties of metals, speculate on how Canadian pioneers made metal rims fit tightly on the wooden wheels of wagons like the one in **Figure 6**.

45. Draw your own home. Redesign it to be as energy-efficient as possible. Make it use little energy in the winter and stay cool in the summer. Label the new features that you would include. Explain how each feature would control heat transfer. What are the costs you would incur?

46. Design an electric kettle so it wastes as little electrical energy as possible.

47. Heating and cooling systems become inefficient when they are not properly maintained. One problem is dust that gathers on heating and cooling coils.
   (a) Why is dust a problem?
   (b) What other maintenance problems can you suggest should be addressed?

48. Design a system that uses garbage to produce electricity for a small community. (If you research "biomass," you will find out more about this method of conserving energy.)

49. The diver in **Figure 7** is wearing a wet suit for protection in cold water. Research and report on the properties of the materials used in wet suits.

Figure 6

Figure 7

50. **Figure 8** shows an astronaut in space, where temperatures in the shade are extremely low and there is no air. Describe ways that the astronaut is protected from the cold, and explain why the suit is light in colour.

Figure 8

51. Outdoor swimming pools lose a lot of heat to the air, especially at night in spring and fall. This heat is often replaced using a heater. Design a cover for a pool to reduce the amount of heat needed.

Glossary

B

boiling point: the temperature at which a substance boils

C

cogeneration: the process of providing electricity and heat at the same time

conduction: the transfer of heat by the collisions of particles in a solid

contraction: a decrease in the volume of an object or substance

convection: the transfer of heat by the movement of particles from one part of a fluid to another

convection current: the motion of the fluid particles during convection

E

expansion: an increase in the volume of an object or substance

G

gas: a substance that fills any container it is in and takes on the shape of the container; it can flow, and it is easy to compress

greenhouse effect: the process of trapping radiant heat inside a structure made with glass

H

heat: energy that is transferred from hotter substances to colder ones

heat capacity: a measure of the amount of heat needed to raise the temperature of the substance or a measure of how much heat the substance releases as it cools

heat conductor: a substance that conducts heat well

heat pollution: wasted heat being produced by human activities

heating curve: a graph of a change of state as a substance is heated

L

lubricant: a substance used to reduce friction

liquid: a substance with a set volume that will take the same shape as the container it is in; it can flow, but it is difficult to compress

M

melting point: the temperature at which a substance melts

P

particle theory: a model that says that all matter is made up of tiny particles that are too small to be seen and are always moving

R

radiant energy: energy transferred by radiation; examples are heat and light

radiation: the transfer of energy by means of waves

S

solid: a substance with a set volume and a rigid shape; it cannot flow and it is very hard to compress

T

thermocouple: a device that uses electricity to measure temperatures

thermometer: a device that uses the expansion and contraction of a liquid to measure

thermostat: a device that uses the expansion and contraction of solids to measure temperatures

temperature: a measure of the average energy of motion of the particles of a substance

Index